TABLE OF CONTENTS

CHAPTER 1: INTRODUCTION 1

CHAPTER 2: CLIMATE CHANGE AND THE ARCTIC 6
    The Shrinking Arctic 8
    Oil 10
    Natural Gas 13
    Maritime Transportation 13
    Northwest Passage 14
    Northeast Passage (or Northern Route) 15
    Central Arctic Passage (or Transpolar Route) 15

CHAPTER 3: CURRENT ARCTIC POLICIES AND CAPABILITIES 17
    Governing Bodies and Policy 17
    National Security Presidential Directive – 66 and Homeland Security Presidential Directive – 25 21
    The U.S. Northern Command 27
    The U.S. Navy 29
    The U.S. Coast Guard 33

CHAPTER 4: UNITED NATIONS CONVENTION ON THE LAW OF THE SEA 40
    UNCLOS Parts and Articles 42
        Freedom of Navigation 42
        Economic Exclusion Zone 43
        Continental Shelf 45
        Commission on the Limits on the Continental Shelf (CLCS) 47
        International Tribunal on the Law of the Sea (ITLOS) and other Dispute Resolution Mechanisms 48
        International Seabed Authority 49

CHAPTER 5: THE UNITED STATES AND UNCLOS 52

CHAPTER 6: U.S. STRATEGIC CONSIDERATIONS IN THE ARCTIC 61
    Russia 61
    China 64
    Canada 66
    Europe 68

CHAPTER 7: CONCLUSION AND RECOMMENDATIONS 71
APPENDIX 75

BIBLIOGRAPHY 79
VITA 90

CHAPTER 1: INTRODUCTION

Climate change is a phenomenon the United States cannot continue to ignore. Global temperatures have been steadily increasing for several decades and physical changes are occurring in almost every ecosystem on the planet. The cause of warming continues to be a source of bitter disagreement between scientists, environmentalists, and politicians. But one thing is certain; the significant loss of ice in the Arctic brought about by this climate change will create serious strategic challenges and opportunities for the United States in the very near future. This thesis will address both sides of the United Nations Convention on the Law of the Sea (UNCLOS) argument, examine current U.S. Arctic strategy and policy (ends), discuss the capabilities of the United States to implement that policy (means), and look at the various mechanisms for implementing policy (ways).

During the past 40 years, Arctic sea ice has thinned by more than 60 percent- from an average thickness of 9 feet to 3 feet.[1] Projections vary, but an ice-free Arctic is becoming a reality and data suggests that the Arctic could experience ice-free summers by 2030. Other more frightening models predict that this could happen as early 2013.[2]

So why should the United States worry about such an event? These upcoming ecological changes could drastically reshape the Arctic's ecosystem creating major social economic, military, and environmental challenges throughout the world. Unfortunately, this is only where the problems begin.

---

[1] Bruce E. Johansen, *Global Warming 101*, (Middletown: Greenwood Press), 39.
[2] David W. Titley and Courtney C. Saint John, "Arctic Security Considerations and the U.S.Navy's Arctic Roadmap," *Arctic Security in an Age of Climate Change Arctic Security in an Age of Climate Change,* (New York: Cambridge University Press, 2011), 36.

As the ice melts, man will have unprecedented access to one of the world's most historically remote regions and to the riches that it contains. These resources include, but are not limited to, fishing, new sea lines of communication for maritime transportation, and harvesting of minerals. The most highly sought out resources, however will be oil and natural gas. In 2008, a study conducted by the United States Geological Survey (USGS) confirmed the Arctic is estimated to hold 30 percent of the world's undiscovered natural gas and 13 percent of the world's undiscovered natural oil.[3] While the global demand for scarce natural resources continuously increases, countries with Arctic borders are making legitimate, and in some cases extravagant, sovereignty claims of Arctic seabeds with the hopes of gaining rights to the untapped riches that the region holds.

Recent and dramatic changes in the Arctic environment are beginning to re-kindle fears that the United States could once again face security challenges in the far North. As part of its effort to create a comprehensive presence in the Arctic, Russia has been steadily expanding its military component there since 2007.[4] Some believe that this may lead the U.S. to face another Cold War-type scenario in the Arctic, this time for resources. Others see signs of cooperation between nations and an opportunity to structure properly the Arctic for the mutual benefit of the international community. The retreat of the Arctic's ice could pose another risk to peace and stability, one that is much more serious than the advent of any "resource war." Countries in the Arctic, including historically strong North Atlantic Treaty Organization (NATO) countries like Canada and Norway, are starting to bicker about navigational rights to important waterways and are staking

---

[3] Geneviève King Ruel, "The (Arctic) show must Go on", *International Journal* 66, no. 4 (autumn 2011), p. 826.
[4] Ariel Cohen, "Russia in the Arctic: Challenges to U.S. energy and Geopolitics in the High North," in *Russia in the Arctic* (Carlisle: Strategic Studies Institute, 2011), 21.

claims to sea lines of communication that may become traversable in the coming years. As the waters of the Arctic Ocean become increasingly more navigable, Russia and the United States may start to feel threatened by the growing presence of foreign governments in areas that they regard as a strategically important exclusive domain.[5]

The United States is once again taking a serious interest in Arctic issues. The United States government agencies and the military, as well as private industry, are moving forward with plans for a myriad of Arctic contingencies that could occur in the very near future. The *National Security Presidential Directive-66/Homeland Security Presidential Directive-25* is the current cornerstone of the United States Arctic policy. This seminal 2009 document acknowledges that changes in the Arctic will affect the nation and affirms that the United States will "meet national security and homeland security needs relevant to the Arctic region."[6] Using this document as a framework for future operations, the Coast Guard and Navy are conducting assessments of their own capabilities to operate in the Arctic. The general consensus is that the military is woefully unequipped to conduct operations in the ice. In the very near future, this could create a serious problem for the nation as the high tempo of world-wide military operations continues and individual programs compete for attention from ever-shrinking budgets.

The United Nations is the obvious forum for conflict resolution between nations stemming from changing conditions in the Arctic. The proposed framework defining the maritime environment for all countries is the United Nations Convention on the Law of

---

[5] Roger Howard, The Arctic Gold Rush : The New Race for Tomorrow's Natural Resources (London; New York: Continuum, 2009), p. 21.
[6] President, Proclamation, "National Security Presidential Directive 66/Homeland Security Presidential Directive 25: Arctic Region Policy," January 9, 2009.

the Sea, or UNCLOS. This document took long-standing customary maritime traditions and codified them into signature law. Additionally, it established many of the international maritime boundaries and created the legal structure so countries could claim exclusive rights to bodies of water and enforce their national laws in a number of areas: fishing, mineral harvesting, pollution, and immigration to name a few. Acting as a steward for the oceans, UNCLOS also created international environmental protection rules and guidelines for natural resource management. Perhaps most importantly, UNCLOS created a mechanism for maritime dispute resolution between countries. UNCLOS has gone through several iterations over the years, with the most current version taking effect in 1994.

Over the last 60 years, the United States has played a large part in contributing to the development of UNCLOS. Although the United States recognizes the tenets of UNCLOS as codified customary maritime law, it is the only Arctic nation that has not formally ratified UNCLOS. Since the early 1980s the Senate has debated the issue and considered ratification of UNCLOS, but has never brought UNCLOS to a full Senate vote. While legitimate arguments for and against ratification of UNCLOS exist, rapidly changing conditions in the Arctic issues are forcing a decision.

While acknowledging that U.S. ratification of UNCLOS has serious implications for a multitude of world-wide political, economic and military issues, this thesis will focus solely on Arctic issues and the strategic environment the U.S. now faces resulting from an ice-free Arctic. It will analyze the national and international plans and policies in place to deal with this emerging challenge, while taking into account the current and historical implications for possible conflict and resolution of disputes.

Finally, this thesis will examine scientific data, history, and the current and projected political state of the Arctic, and submit conclusions in support of a recommendation that strategic considerations and national interests brought about by changes in the Arctic now require the United States to ratify the United Nations Convention on the Law of the Sea (UNCLOS).

CHAPTER 2: CLIMATE CHANGE AND THE ARCTIC

The debate about climate change is not new. Climatologists have recorded and studied rising temperatures for more than a century. Since the 1950s, hundreds of climate change studies and theories have been promoted, with the possible effects ranging from minor change to doomsday scenarios claiming extinction. While most generally accept that global temperatures are on the rise, the explanations for this phenomenon vary. As with nearly every issue in our modern society, it is impossible to exclude politics from the argument. Politicians in America use climate change either as a rallying cry for radical change or try to debunk it as political theater. *National Journal* magazine conducted a poll in 2007 and asked the question "Do you think it's been proven beyond a reasonable doubt that the Earth is warming because of man-made problems?" In response, 95 percent of self-identified Democrats said yes while only 13 percent of self-identified Republicans said yes.[1] Regardless of political outlook ice thaw in the Arctic is a reality, and more importantly, an emerging strategic challenge.

To frame the issue, one needs to understand the baseline elements in the atmosphere. The Earth's atmosphere is made up of two basic gases, consisting of about 78 percent nitrogen and 21 percent oxygen. The remaining one percent is made up of a mix of carbon dioxide, methane, argon, and ozone.[2] The "greenhouse effect" is when heat gets trapped in the atmosphere. This is a natural function and necessary for temperatures to stay warm enough to sustain life. Carbon dioxide, as well as methane

---

[1] Richard E. Cohen and Peter Bell, "Congressional Insiders Poll: Do You Think it's been Proven Beyond a Reasonable Doub that the Earth is Warming because of Man made Problems?" *National Journal*, (February 3, 2007), p. 6. 72 people (41 Democrats and 31 Republicans) were asked the question. The possible answers to the "man-made" question were: yes, no, and only part of the cause.

[2] Bruce E. Johansen, *Global Warming 101*, (Middletown: Greenwood Press), 1.

and other greenhouse type gases, keeps heat locked in the atmosphere. As the atmosphere is exposed to ever increasing levels of carbon dioxide, it traps more heat and ambient temperatures rise. The scientific world believes that the primary culprits in excessive carbon dioxide creation are man-made and that "modern climate change is dominated by human influences, which are now large enough to exceed the bounds of natural variability . . . . Anthropogenic climate change is now likely to continue for many centuries."[3] In the summer of 2003, Europe experienced an unprecedented heat wave. Over 35,000 people died because of scorching temperatures. London had a recorded temperature of 100 degrees, the first in over 400 years of recorded weather data. If scientific models hold, by the 2040s one out of every two summers in Europe will be hotter than 2003.[4] By the end of the century, projections indicate that north Florida will have more than 165 days (nearly six months) per year over 90°F, up from roughly 60 days in the 1960s and 1970s.[5] Summer high temperatures in the United States continue to break new records. July 2012 is a case in point, distinguishing itself as the hottest month recorded in United States history with an average temperature of 77.6°F. That is 3.3°F higher on average than anytime during the twentieth century. During this month, 62.9 percent of the contiguous United States was experiencing drought conditions.[6] Along with rising temperatures, experts say climate change is also seen as the primary

---

[3] Thomas Karl and Kevin Trenberth, "Modern Global Climate Change," *Science Vol 202, no. 5651*, (2003): 1719.

[4] Bruce E. Johansen, *Global Warming 101*, (Middletown: Greenwood Press), 18.

[5] The United States Global Change Research Program, Committee on Environment and Natural Resources, http://downloads.globalchange.gov/usimpacts/pdfs/southeast.pdf, (Accessed September 07, 2012).

[6] National Oceans and Atmospheric Administration, "Climate Watch Magazine: Hottest Month Ever Recorded," NOAA, http://www.climatewatch.noaa.gov/image/2012/july-2012-hottest-month-on-record (Accessed September 17, 2012).

reason for rising sea levels, rising sea temperatures, and melting ice in the Arctic and Antarctic regions.

The Shrinking Arctic

The Artic is below the constellation Ursa Minor, or the Great Bear. Hence, the word Arctic comes from the Greek word *arktos* or "bear."[7] The generally accepted definition of the Arctic is of the Earth North of latitude 66° 33´N, better known as the Arctic Circle. Since Robert Peary first discovered the North Pole in 1909, the Arctic has been a source of fascination for people. The aspect of the Arctic that makes it so unique to Earth is the regional climate and topography. A desolate place north of the tree line, the average temperatures in the Arctic range from the mid-40°F in the summer to the negative mid-20°F in the winter. Over the decades, American interest in the Arctic has waxed and waned in conjunction with a myriad of economic booms and political tempests. The Arctic is starting to gain American and international attention once again because of climate change. One of the most noticeable symptoms of climate change is the ever decreasing amount of summer ice in the Arctic. Arctic ice is melting at rates never seen before in recorded history. Between the years 2002 and 2006, more ice melted than during any period before. It is important to note that when an "ice free" Arctic is discussed, this means exclusively ice free during short periods in the summer even though significant sea ice remains throughout the Arctic for the remainder of the year. The ice free summer period may be extended in coming years. Using the collected data in 2006, researchers say that the Arctic could be ice free in the summer as soon as 2070. The very next year in 2007, the Arctic experienced the single largest yearly ice

---

[7] Shelagh D. Grant, *Polar Imperative* (Vancouver, BC: Douglas & McIntyre, 2010), 5.

melt in history. Soon after, researchers moved the ice free target up by 30 years.[8] As of September 16, 2012 the extent of the Arctic ice fell below 1.32 million square miles. This was 300,000 square miles (as a reference, the state of Texas is 268,600 square miles) below the previous record low of 1.61 million square miles set in September 2007.[9] In September of 2012 using satellite data, the Canadian Ice Service found that just 12 percent of the region was frozen, compared with a normal 30 percent to 35 percent.[10] Summer thaw in the Arctic is happening and ice free summers have become a very real possibility. Some academics have drawn historical comparisons between the Arctic Ocean and the Mediterranean Sea. As one accomplished Canadian academic said: "What the Aegean Sea was to antiquity, what the Mediterranean was to the Roman world, what the Atlantic Ocean was to the expanding Europe of the Renaissance days, the Arctic Ocean is becoming to the world of aircraft and atomic power."[11] Some even go so far as to say that the "age of the Arctic" is upon us. Some believe that access to the Arctic could destabilize security in the region and spark a resource war among Arctic nations. The question at hand is: how will America balance the competing interests and demands that will be unleashed by an ice-free Arctic? American leadership must understand the significance of the ice free summer on oil and gas exploration and maritime commerce in the Arctic.

---

[8] Bruce E. Johansen, *Global Warming 101*, (Middletown: Greenwood Press), 41.
[9] National Aeronautical and Space Administration, "Arctic Sea Ice Hits Smallest Extenst in Satellite Era," NASA, http://www.nasa.gov/topics/earth/features/2012-seaicemin.html (accessed January 13, 2013).
[10] Canadian Press, "Summer Takes Uprecedented Toll on Arctic Ice," *canada.com*, September 19, 2012. http://o.canada.com/2012/09/19/summer-takes-unprecedented-toll-on-arctic-ice-prompting-global-warming-fears/ (accessed September 30, 2012).
[11] Shelagh D. Grant, *Sovereignty or Security? Government Policy in the Canadian North, 1936-1950*, (Vancouver: Univ of British Columbia Press, 1988), p. 210.

Oil

With oil supplies dwindling for some countries and the demand for energy exponentially increasing world-wide, the international community is hoping that the Arctic holds treasures that could sustain energy consumption requirements for generations to come. First and foremost on the agenda of Arctic nations is the enticing prospect of gaining access to the large oil fields many geologists believe exist below the waters and tundra of the Arctic Circle. While oil estimates can be difficult to make, geologists that have studied the Arctic say that as much as 20 percent to 25 percent of the world's undiscovered recoverable oil could be hidden in the underwater seabeds of the Arctic. A U.S. Geological Survey released in July of 2008 estimated that a staggering 90 billion barrels of recoverable oil lie north of the Arctic Circle.[12] According to geologist David Gautier, who led the study, the Alaskan Arctic Province, belonging to the United States, holds the most amount of undiscovered, untapped oil at approximately 30 billion barrels. "In our judgment," the study concluded, "the Arctic Alaskan Province is the most obvious place to look for oil north of the Arctic Circle right now."[13]

Large oil companies have been eying the Arctic for decades in conjunction with oil speculation; however, excessive costs and the natural barriers of the region historically have precluded drilling. Costs of onshore oil drilling and recovery in Alaska are 50 percent to 100 percent more expensive than a site in Texas.[14] American and

---

[12] United States Geological Survey, "Circum-Arctic Resource Appraisal: Estimates of Undiscovered Oil and Gas North of the Arctic Circle." United States Department of the Interior, http://pubs.usgs.gov/fs/2008/3049/fs2008-3049.pdf (accessed September 10, 2012).

[13] Queenie Wong, "New Study Estimates Vast Supplies of Arctic Oil , Gas," *McClatchy Newspapers*, July 24, 2008, 2008, http://www.mcclatchydc.com/2008/07/23/v-print/45349/new-study-estimates-vast-supplies.html (accessed September 12, 2012).

[14] United States Energy Information Administration, "Arctic Oil an Natural Gas Resources," United States Department of Energy, http://www.eia.gov/oiaf/analysispaper/arctic/index.html#adcr (accessed September 20, 2012).

international energy companies are wasting no time in preparing for eventual access to this region in hopes of major oil finds. Energy companies already have identified more than 400 possible oil and natural gas sites inside the Arctic Circle.

In a plan that has spanned over three years, Shell Oil has worked diligently with federal and state officials to gain access to proposed sites in American waters north of the Arctic Circle. No stranger to Alaska, Shell Oil was one of the first companies to beginning producing oil through platforms in Cook Inlet during the 1960s. In the 1990s, Shell looked at drilling in the Arctic waters, but decided that it was cost prohibitive. In September of 2012, Royal Dutch Shell received permission from the U.S. Bureau of Safety and Environmental Enforcement (BSSE) to conduct preparatory drilling in the Chukchi and Beaufort Seas. Shell's Alaska exploration manager Steve Phelps said, "This is the stuff that most of the world was finding in the 1930s, the 1950s, the 1960s, in places like Saudi Arabia and the Middle East, Nigeria. This one potential resource far outweighs any single field we've got in the Americas' portfolio."[15] The process cost Shell over $4.5 billion for the purchasing of Arctic leases; Shell was required to create extensive plans to convince the U.S. government that emergency procedures were in place to handle the litany of environmental issues associated with drilling for oil. Senator Lisa Murkowski of Alaska was quoted as saying "this represents great news for Alaska and the entire country. I cannot overstate the opportunity that Arctic exploration offers in terms of jobs and energy security."[16]

---

[15] Lisa Demer, "Shell Gambles Billions in Arctic Alaska Push," *Anchorage Daily News*, December 04, 2011.
[16] Nick Snow, "BSEE Approves Shell's Chukchi Sea Oil Spill Response Plan," *Oil & Gas Journal*, 110, 2C, (Feb 27, 2012): 14 – 15.

Senator Murkowski's sentiments are not shared by everyone. Drilling in the Arctic has always been a contentious issue. Alaskan Native groups indigenous to the North Slope are rightfully wary of the potential problems that industry could bring to their societies and the environment that they depend upon to survive. Native groups have banded together and filed multiple lawsuits attempting to stop or at least stall Shell's drilling. In addition, environmental groups like Greenpeace have staged large scale protests and conducted heavy public relations campaigns to stop Shell. They argue that disasters similar to *EXXON VALDEZ* and *DEEPWATER HORIZON* could just as easily happen in the Arctic and that Shell has shown little ability to manage such a crisis. Royal Dutch Shell has had several setbacks, including damage to safety and emergency response equipment that quickly shut down exploratory drilling that is not expected to start up again until 2013.

Not only are activist and indigenous groups asking questions about safety, several high profile governments are looking at the possibility that the technology and safety procedures currently do not exist to handle a disaster contingency adequately. In late September 2012, the Environmental Audit Committee of the British House of Commons called for a complete halt to all international Arctic drilling until major safeguards and financial guarantees could be ensured. The chairwoman of the committee, Joan Walley, summed up the argument by saying "the infrastructure to mount a big clean-up operation is simply not in place and conventional oil spill response techniques have not been proven to work in such severe conditions."[17]

---

[17] David Stringer, "UK Lawmakers Seek Moratorium on Arctic Drilling," *Seattle Times*, (September 19, 2012).

Natural Gas

Natural gas is seen as a cleaner alternative to oil that produces no solid waste and is gaining momentum as a popular and more efficient form of energy in the United States. Already an accepted energy source for mass transportation across the country, natural gas is beginning to challenge coal as the number one fuel for power production. In 2012 alone, the demand for natural gas used in conjunction with electricity creation was up 24 percent from the previous year. This sharp rise in demand can be explained by its relative inexpensive cost compared to traditional oil and its emergence as a partial replacement for coal powered electrical plants.[18]

While the Arctic area also holds large areas of natural gas, there are unique economic and technological impediments that would make the development of natural gas resources very challenging. The expense of developing and transporting natural gas can quickly outweigh profit. The extremely high cost of extraction and transportation makes it virtually certain that only the most powerful global energy companies will be able to undertake a development of natural gas resources.

Maritime Transportation

As the ice recedes in the Arctic, the benefits of using the newly navigable waters are clear, and trans-continental shipping companies are actively making plans to use these routes to save time and mileage. Three Arctic Circle routes are being examined for use: the Northwest Passage, the Northeast Passage (or Northern Route), and the Central Arctic (or Transpolar) shipping route.

---

[18] United States Energy Information Administration, "Natural Gas Demand at Power Plants was High in Summer 2012," United States Department of Energy, http://www.eia.gov/todayinenergy/detail.cfm?id=7870 (accessed January 31, 2013).

In 2009, the U.S. Arctic Research Commission (USARC) released a four-year study on shipping trends in the Arctic. The Arctic Marine Shipping Assessment looked at what types of vessels are operating in the Arctic and at what frequency. The types of vessels using the Arctic were put into one of four major groups: fishing, tourism, community re-supply, and bulk carrier. Over 50 percent of the 6,000 vessels were fishing vessels, but a substantial number were bulk carriers of oil, gas, and minerals. Of additional interest, the study found that the number of cruise ships that made port calls in Greenland between 2006 and 2007 went up from 157 to 222.[19] This analysis will look briefly at the three major Arctic sealines of communication and discuss them in the context of an ice-free Arctic.

Northwest Passage

Originally, named the Strait of Anian by the Spanish, this waterway was thought to be the quickest way from the North Atlantic to the Pacific. This waterway could allow ships to enter the Arctic between Canada and Greenland, transit through the archipelago islands of Canada, and then exit through the Beaufort Sea and Bering Sea into the North Pacific. The Northwest Passage could become a more attractive option for marine transportation as waterways become less icy and remain so for longer periods. For example, a marine transit from Rotterdam to Shanghai utilizing the Northwest Passage would be approximately 9,297 thousand miles vice 12,107 thousand miles when transiting through the Suez Canal.[20] During the summer of 2007, the Northwest Passage

---

[19] United States Arctic Research Commission, "Arctic Marine Shipping Assessment: Current Marine use and the AMSA Database," United States Arctic Research Commission, http://www.arctic.gov/publications/AMSA/current_marine_use.pdf (accessed September 10, 2012).p. 279

[20] Paul Arthur Berkman and Royal United Services Institute for Defence and Security Studies., *Environmental Security in the Arctic Ocean : Promoting Co-Operation and Preventing Conflict* Abingdon:

became completely navigable to maritime traffic for the first time in history. By offering savings in cost, time, and mileage, predictions are that the Northwest Passage will become a major trans-continental route for oil, natural gas, and bulk carriers.

Northeast Passage (or Northern Route)

The Northeast Passage, or Northern Sea Route, runs from the North Sea in the Atlantic, following across the top of Russia, ending in the Bering Sea in the Pacific. The first ships to transit the length of the Northeast Passage without icebreaker assistance did so in 2009. The transit from Rotterdam to Yokohama, Japan through the Northeast Passage is 4,450 miles shorter than the current Suez Canal transit. In 2010, a Norwegian shipping company took the voyage from Norway to China in 21 days compared with the 37 days typically needed to use the Suez Canal. Estimates are that this shorter transit saved over $300,000 a trip.[21] In 2011, 18 additional ships made the nearly ice-free transit.[22] The Northeast Passage is still a dangerous transit due to weather and ice, but as the summers become more ice-free, it is predicted that the number of transiting ships will increase exponentially.

Central Arctic Passage (or Transpolar Route)

Receding ice around the North Pole could eventually open up the shipping route that is becoming known as the Central Arctic Passage or the Transpolar Route. In July of 2012, a Chinese icebreaker, escorted by a Russian nuclear icebreaker, made the dangerous transit across the North Pole. Until recently this passage was only theoretical,

---

published on behalf of The Royal United Services Institute for Defence and Security Studies by Routledge Journals, 2010), p. 71.

[21] Andrew E. Kramer, "Warming Revives Dream of Sea Route in Russian Arctic," *New York Times*, October 17, 2011.

[22] Arctic Portal, "Shipping," Nordurslodagattin, Akureyri, Iceland, http://www.arcticportal.org (accessed September 26, 2012).

but with the loss of summer ice, a commercial transit across the pole could become a common event in the near future. Shipping companies involved with transcontinental maritime commerce could significantly reduce the transit distance from Murmansk in the Atlantic to ports south of the Bering Strait in the Pacific.

The Arctic is once again gaining in strategic importance. Arctic nations and nations with economic ties to international maritime shipping are scrambling to create policy and build capability to position themselves as Arctic players. The U.S. is slowly learning that it lacks both executable policy and capability to operate in the Arctic; however, renewed importance has been placed on the region by U.S. Arctic stakeholders. Becoming a ratified party to UNCLOS could be a major focus of future U.S. Arctic efforts.

CHAPTER 3: CURRENT ARCTIC POLICIES AND CAPABILITIES

Responsibility for the Arctic, nationally and internationally, is not something that can be explained easily. There are literally dozens of private and public research, activist, and policy organizations that work on Arctic issues. Sometimes cooperatively, sometimes independently, and sometimes with competing interests, these groups all vie to have their positions heard at different levels of governance. On the international stage, Arctic countries work together on cooperative councils that may or may not have binding resolutions. Although Arctic countries may understandably be looking to protect national interests and capitalize on exploitation of resources, a spirit of cooperation and goodwill has so far prevailed, reflected in the number of international meetings and councils between Arctic nations.

The following chapter will be divided into three sections. First, it will examine the historical and most current strategic documents dictating modern U.S. Arctic policy. Secondly, it will discuss the relevant international and U.S. level bodies that govern Arctic issues. Finally, it will analyze current U.S. military initiatives and strategy documents to identify challenges in capabilities shortfalls in implementing national Arctic strategy.

Governing Bodies and Policy

The United States became an Arctic nation in 1867 with the purchase of Alaska from Russia. Since that time, the United States has had limited policy or doctrine with respect to the Arctic. That all changed with the challenges associated with the Cold War with the Soviet Union. The dangerous proximity of United States military assets to

Soviet assets, led President Nixon to establish the first modern U.S. Arctic policy in the form of *National Security Decision Memorandum (NSDM) 144*. This document recognized the strategic and environmental importance of the Arctic and stressed three overarching principles: national security, environmental stewardship, and international cooperation.[1] Outside of the Cold War, the Arctic received very little attention until the environmental movement gained influence in the public discourse. The Reagan administration created the *Arctic Research and Policy Act of 1984*. While not ignoring the fact that the United States shared an international maritime boundary with the Soviet Union in the Arctic, the document outlined several key tenets on Arctic research and environmental protection. Its two biggest accomplishments were the establishment of the Arctic Research Commission (ARC) and the Interagency Arctic Research Policy Committee (IARPC).[2] Although not policymaking bodies, both of these important entities were charged with conducting research and making recommendations to the current administration on scientific matters dealing with the Arctic.

Under the Bush administration, the United States joined the other Arctic nations and several indigenous groups in 1991 in signing the *Arctic Environmental Protection Strategy*. This non-binding cooperative agreement looked at the Arctic through a stewardship focus to prevent pollution and to respond to emergencies. Although a positive step forward in the post-Cold War era for cooperation between countries, this initiative provided no legal authority and lacked any real direction.

---

[1] Henry Kissinger, "National Security Decision Memorandum 144," Federation of American Scientists, http://www.fas.org/irp/offdocs/nsdm-nixon/nsdm-144.pdf (accessed October 4, 2012).

[2] United States Congress, *Arctic Research and Policy Act of 1984 (Amended 1990)*, Vol. Public Law 98-373; Public Law 101-609 (Washington, DC, 1984; 1990).

In 1996, during an international meeting in Canada that came to be known as the *Ottawa Declaration*, the eight Arctic nations (United States, Russia, Canada, Denmark, Finland, Sweden, Iceland, and Norway) created the most important Arctic governing body to date. The Arctic Council was a high-level governing body established to assure a dialogue between the Arctic nations, as well as other interested countries and indigenous peoples of the Arctic Circle. The group meets every six months to discuss the progress of its six permanent working groups. These working groups address topics ranging from sustainment of natural resources to environmental and animal protection and pollution issues. Significantly, the Arctic Council does not work on military matters.

At the end of every two years, the group names a council member nation to serve as secretariat; it produces a declaration report, typically named after the city that hosts the forum, outlining the progress made in the previous two years and define the way-ahead for the group in the next two years. The United States is scheduled to head the group for 2015 – 2017. The United States views the Arctic Council as the most useful forum for discussing international Arctic issues outside of the United Nations. For example, Secretary of State Hillary Clinton along with senior members of the Arctic Council signed the first binding international agreement between the Arctic nations that broke the Arctic down into regions and assigned areas of responsibility to each Arctic Council nation for search and rescue.[3] Critics are quick to point out the lack of tangible results for Arctic governance. They point to the fact that changes in the Arctic will inevitably cause friction in the region and that the Arctic Council simply is not set up to handle dispute resolution.

---

[3] North American Aerospace Defense Command and United States Northern Command, Arctic Collaborative Workshop: Arctic Oil Spill & Mass Rescue Operation – Tabletop Exercise, After Action Report (Colorado Springs: USNORTHCOM, 2012), p. 29.

One bright spot of international cooperation involving the Arctic Council was the Arctic Search and Rescue Exercise (SAREX) of 2011 sponsored by USNORTHCOM. The first of its kind, all eight Arctic nations supplied rescue capabilities that worked cooperatively off the coast of Greenland to assist in a cruise ship disaster scenario. Touted as a major international success, the SAREX highlighted the daunting difficulties that exist in mounting search and rescue efforts in the Arctic.

A notable meeting called the Ilulissat Declaration outside of the Arctic Council took place in Greenland in 2008. The five littoral Arctic nations, without Finland, Iceland, or Sweden, met to discuss the way ahead for policies that should govern activities in the Arctic. The declaration stated "the law of the sea provides for important rights and obligations concerning the delineation of the outer limits of the continental shelf, the protection of the marine environment, including ice-covered areas, freedom of navigation, marine scientific research, and other uses of the sea . . . . We therefore see no need to develop a new comprehensive international legal regime to govern the Arctic Ocean."[4] In sharp contrast from the type of hard codified international law that governs Antarctica, the *Ilulissat Declaration* of 2008 prevents law making bodies from limiting the international community in its pursuit of sovereignty and resources. On face value, this agreement was lauded for easing international tensions; however, it can be argued as many experts do, that the agreement was actually counterproductive and put the health of the region, as well as regional security, at significant risk. Of particular note, the pinnacle point of the declaration commits the Arctic nations to abide by the United Nations Convention on the Law of the Sea (UNCLOS).

---

[4] Ilullisat Conference, *The Ilullisat Declaration* (Ilullisat, Greenland, 2008), 1-2.

Currently, Arctic policy within the United States is handled by the State Department's Office of Ocean and Polar Affairs (OPA), which is largely responsible for day to day Arctic issues for the U.S. government. The U.S. Arctic Policy Group (APG), however, is the real power. Headed by the Secretary of State, the APG is an interagency group that meets once a month with the President.

National Security Presidential Directive – 66 and Homeland Security Presidential Directive – 25

All U.S. security policy can be linked back to support one or more of the four enduring national interests listed in the President's *National Security Strategy* (NSS). Those enduring national interests are as follows:

1. Security: The security of the United States, its citizens, and U.S. allies and partners.
2. Prosperity: A strong, innovative, and growing U.S. economy in an open international economic system that promotes opportunity and prosperity.
3. Values: Respect for universal values at home and around the world.
4. International Order: An international order advanced by U.S. leadership that promotes peace, security, and opportunity through stronger cooperation to meet global challenges.[5]

In an improvement from previous national security strategies, the 2010 *NSS* specifically addresses the Arctic by saying, "the United States is an Arctic nation with broad and fundamental interests in the Arctic region, where we seek to meet our national security needs, protect the environment, responsibly manage resources, account for indigenous communities, support scientific research, and strengthen international cooperation on a wide range of issues." The *NSS* goes a step further by calling for U.S. ratification of the United Nations Convention on the Law of the Sea (UNCLOS).[6]

---

[5] U.S. President, *National Security Strategy* (Washington, DC: United States, 2010).
[6] Ibid.

As part of fully realizing its enduring national interests in the Arctic, President George W. Bush authored *National Security Presidential Directive (NSPD) - 66* and *Homeland Security Presidential Directive (HSPD) – 25, Arctic Region Policy* in 2009. This document is the current cornerstone for U.S. Arctic policy. It is important to note that while these are actually two separate documents, they both contain verbatim the same information and policy mandates. The policy outlines six broad policy goals as they pertain to various emerging Arctic issues, placing a heavy emphasis on security, international cooperation, sustainment and management of natural resources, and environmental stewardship of the Arctic. The six stated policy goals for the U.S. in *NSPD-66/HSPD-25* are as listed:

1. Meet national security and homeland security needs relevant to the Arctic region;
2. Protect the Arctic environment and conserve its biological resources;
3. Ensure that natural resource management and economic development in the region are environmentally sustainable;
4. Strengthen institutions for cooperation among the eight Arctic nations (the United States, Canada, Denmark, Finland, Iceland, Norway, the Russian Federation, and Sweden);
5. Involve the Arctic's indigenous communities in decisions that affect them;
6. And enhance scientific monitoring and research into local, regional, and global environmental issues.[7]

In addition to the policy goals, *NSPD-66/HSPD-25* identifies seven areas of focus that support those policy goals set for the U.S. in the Arctic. These focus areas include national security, international governance, extended continental shelf and boundary issues, international scientific cooperation, maritime transportation, economic issues (to include energy), and environmental protection. For each of these focus areas, there are stated issues, assumptions, and directive guidance on implementation. As an example,

---

[7] President. Proclamation, "National Security Presidential Directive 66/Homeland Security Presidential Directive 25: Arctic Region Policy", (January 9, 2009).

NSPD-66/HSPD-25 identifies "national security and homeland security interests in the Arctic" as one of its focus areas.

1. The United States has broad and fundamental national security interests in the Arctic region and is prepared to operate either independently or in conjunction with other states to safeguard these interests. These interests include such matters as missile defense and early warning; deployment of sea and air systems for strategic sealift, strategic deterrence, maritime presence, and maritime security operations; and ensuring freedom of navigation and overflight.

2. The United States also has fundamental homeland security interests in preventing terrorist attacks and mitigating those criminal or hostile acts that could increase the United States vulnerability to terrorism in the Arctic region.

3. The Arctic region is primarily a maritime domain; as such, existing policies and authorities relating to maritime areas continue to apply, including those relating to law enforcement. Human activity in the Arctic region is increasing and is projected to increase further in coming years. This requires the United States to assert a more active and influential national presence to protect its Arctic interests and to project sea power throughout the region.

4. The United States exercises authority in accordance with lawful claims of United States sovereignty, sovereign rights, and jurisdiction in the Arctic region, including sovereignty within the territorial sea, sovereign rights and jurisdiction within the United States exclusive economic zone and on the continental shelf, and appropriate control in the United States contiguous zone.

5. Freedom of the seas is a top national priority. The Northwest Passage is a strait used for international navigation, and the Northern Sea Route includes straits used for international navigation; the regime of transit passage applies to passage through those straits. Preserving the rights and duties relating to navigation and overflight in the Arctic region supports our ability to exercise these rights throughout the world, including through strategic straits.

6. Implementation: In carrying out this policy as it relates to national security and homeland security interests in the Arctic, the Secretaries of State, Defense, and Homeland Security, in coordination with heads of other relevant executive departments and agencies, shall:

a. Develop greater capabilities and capacity, as necessary, to protect United States air, land, and sea borders in the Arctic region;

b. Increase Arctic maritime domain awareness in order to protect maritime commerce, critical infrastructure, and key resources;

c. Preserve the global mobility of United States military and civilian vessels and aircraft throughout the Arctic region;

d. Project a sovereign United States maritime presence in the Arctic in support of essential United States interests; and

e. Encourage the peaceful resolution of disputes in the Arctic region.[8]

As with other national policy documents, the goals of *NSPD-66/HSPD-25* provides linkage back to higher strategic guidance. In this example, the national security interests in the Arctic as listed in *NSPD-66/HSPD-25* are informed by and mirror the enduring national interest stated in the NSS.

Another important focus area identified in *HSPD-66/NSPD-25* is international governance of the Arctic. The following are the objectives and implementation directives for international governance as listed in *HSPD-66/NSPD-25*:

1. The United States participates in a variety of fora, international organizations, and bilateral contacts that promote United States interests in the Arctic. These include the Arctic Council, the International Maritime Organization (IMO), wildlife conservation and management agreements, and many other mechanisms. As the Arctic changes and human activity in the region increases, the United States and other governments should consider, as appropriate, new international arrangements or enhancements to existing arrangements.

2. The Arctic Council has produced positive results for the United States by working within its limited mandate of environmental protection and sustainable development. Its subsidiary bodies, with help from many United States agencies, have developed and undertaken projects on a wide range of topics. The Council also provides a beneficial venue for interaction with indigenous groups. It is the position of the United States that the Arctic Council should remain a high-level forum devoted to issues

---

[8] Ibid.

within its current mandate and not be transformed into a formal international organization, particularly one with assessed contributions. The United States is nevertheless open to updating the structure of the Council, including consolidation of, or making operational changes to, its subsidiary bodies, to the extent such changes can clearly improve the Council's work and are consistent with the general mandate of the Council.

3. The geopolitical circumstances of the Arctic region differ sufficiently from those of the Antarctic region such that an "Arctic Treaty" of broad scope -- along the lines of the Antarctic Treaty -- is not appropriate or necessary.

4. The Senate should act favorably on U.S. accession to the U.N. Convention on the Law of the Sea promptly, to protect and advance U.S. interests, including with respect to the Arctic. Joining will serve the national security interests of the United States, including the maritime mobility of our Armed Forces worldwide. It will secure U.S. sovereign rights over extensive marine areas, including the valuable natural resources they contain. Accession will promote U.S. interests in the environmental health of the oceans. And it will give the United States a seat at the table when the rights that are vital to our interests are debated and interpreted.

5. Implementation: In carrying out this policy as it relates to international governance, the Secretary of State, in coordination with heads of other relevant executive departments and agencies, shall:

a. Continue to cooperate with other countries on Arctic issues through the United Nations (U.N.) and its specialized agencies, as well as through treaties such as the U.N. Framework Convention on Climate Change, the Convention on International Trade in Endangered Species of Wild Fauna and Flora, the Convention on Long Range Transboundary Air Pollution and its protocols, and the Montreal Protocol on Substances that Deplete the Ozone Layer;

b. Consider, as appropriate, new or enhanced international arrangements for the Arctic to address issues likely to arise from expected increases in human activity in that region, including shipping, local development and subsistence, exploitation of living marine resources, development of energy and other resources, and tourism;

c. Review Arctic Council policy recommendations developed within the ambit of the Council's scientific reviews and ensure the policy recommendations are subject to review by Arctic governments; and

d. Continue to seek advice and consent of the United States Senate to accede to the 1982 Law of the Sea Convention.[9]

The international governance focus of *NSPD-66/HSPD-25* ties into the *NSS*'s enduring national interest of promoting a just and sustainable world order. Throughout these various national-level policy documents, it is clear that the U.S. intends to shape the strategic environment by engaging with the international community as a leader and partner of choice. Perhaps the most important part of the international governance portion of *NSPD-66/HSPD-25* is the recognition of the importance for the Senate to work towards full ratification of UNCLOS. In doing so, President Bush made it very clear that accession into UNCLOS was an important part of the United States' Arctic strategy and a vital component to achievement of the Arctic goals outlined in *NSPD-66/HSPD-25*.

United States Arctic policy, in the form of *NSPD-66/HSPD-25*, is a positive first step in recognizing the strategic importance of the Arctic and addressing issues the U.S. will face in the near future. As with all of the aforementioned documents, current national policy is clear that international engagement, not isolationism or unilaterism, will be the norm for protecting and advancing U.S. interests internationally as well as in the Arctic. However, the ambiguous nature and relatively shallow guidance of *NSPD-66/HSPD-25* will continue to hamper efforts by Arctic stakeholders to execute various business, military and governance agendas. Continued reliance on customary law, coupled with a wait-and-see posture, will negate any attempt to clarify U.S. strategic Arctic policy. Ratifying UNCLOS and applying its legal framework to the Arctic would give the U.S. the legal certainty required to create a comprehensive and sustainable Arctic policy for the future.

---

[9] Ibid.

The execution of U.S. national policy in the maritime Arctic falls into several departments, namely the Department of Defense (DOD) and Department of Homeland Security (DHS). Reinforcing the President's position on the importance of the Arctic, the Department of Defense emphasized the necessity for Arctic strategic partnerships in the *2010 Quadrennial Defense Review* (QDR) and stated that "to support cooperative engagement in the Arctic, DOD strongly supports accession to the United Nations Convention on the Law of the Sea."[10] More recently, in a 2012 response to the President's guidance to the Department of Defense, DOD released *Sustaining U.S. Global Leadership: Priorities for a 21st Century Defense* that stated, "the United States will continue to lead global efforts with capable allies and partners to assure access to and use of the global commons, both by strengthening international norms of responsible behavior and by maintaining relevant and interoperable military capabilities."[11] Within DOD and DHS, United States Northern Command, the U.S. Navy, and U.S. Coast Guard are three of the primary U.S. stakeholders that will develop strategy and build capacity to implement national policy in the Arctic.

The U.S. Northern Command

Since its creation in 2002, United States Northern Command (USNORTHCOM) had shared responsibility for the Arctic with two other combatant commanders: United States European Command (USEUCOM) and United States Pacific Command (USPACOM). The April 2011 update to the Unified Command Plan (UCP) reduced the number of combatant commanders responsible for the Arctic to USNORTHCOM and

---

[10] U.S. Department of Defense, *Quadrennial Defense Review Report*, (Washington DC: Department of Defense, February 2010), p. 86.

[11] U.S. Department of Defense, *Sustaining U.S. Global Leadership: Priorities for 21st Century Defense*, (Washington DC: Department of Defense, January 2012), p. 3.

USEUCOM. USNORTHCOM was named as the primary DOD advocate for Arctic issues, but coordination between combatant commanders could become necessary in the case of a crisis. While not explicitly addressed, it could be surmised that USEUCOM retained some responsibility because of its relationship and proximity to Russia. Alaska Command is a sub-unified command subordinate to USPACOM, and USPACOM owns the forces that operate in Alaska. Since there is no Joint Operating Area (JOA) defined in the Arctic outside of the terrestrial boundaries of Alaska, the waters in the Arctic are owned by the naval components to USNORTHCOM and USEUCOM: U.S. Fleet Forces and U.S. Naval Forces Europe, respectively. An operational issue within the land and airspace boundaries of Alaska is managed by standing up Joint Task Force Alaska (JTF-AK), who is subordinate to USNORTHCOM. During normal operations, U.S. Alaska Command (ALCOM), subordinate to USPACOM, is the command that has most experience and access to the Arctic; however, JTF-AK has the assigned mission to "deter, detect, prevent and defeat threats within the Alaska Joint Operations Area . . . in order to protect U.S. territory, citizens, and interests and as directed, conduct civil support."[12] Increased activity in the Arctic may necessitate a new command and control structure to effectively manage U.S. operations.

To assist with USNORTHCOM's new mission, the Arctic Capabilities Assessment Working Group (ACAWG) was formed. This working group examined existing doctrine and capabilities to identify assets needed in the coming years. The ACAWG recently described major challenges and shortfalls:

1. The harsh and challenging environment;
2. Extreme distances between operating areas and support bases;

---

[12] United States Northern Command, "Joint Task Force Alaska," *United States Northern Command*, http://www northcom.mil/About/index html#JTFAK (accessed December 04, 2012).

3. Poor reliability of communications in the northern latitudes;
4. Inadequate situational awareness resulting from limited and or degraded sensing capabilities;
5. Inadequate air and surface asset capability particularly with respect to operation in and on the edge of ice for safety and security missions; and
6. Lack of a logistics infrastructure to support all operations, especially operations of national significance.[13]

The U.S. Navy

The Navy has a long and sometimes tense history in the Arctic. During the Cold War, the U.S. and Canada positioned military forces in the Arctic to intercept Soviet bombers as they flew over the North Pole towards North America. In 1954, *USS NAUTILUS,* the world's first nuclear submarine, traversed the entire Arctic Ocean including the North Pole completely submerged. Five years later, *USS SKATE* completed the same voyage but surfaced at the North Pole for the first time.[14] As the technology and efficiency of nuclear weapons increased in the 1970s, American submarines were capable of firing their weapons from anywhere in the Arctic. At the height of the cold war, the Navy had nuclear ballistic missile submarines deployed to the Arctic continuously. When the Cold War ended, the Navy reduced its presence in the Arctic.

Two cornerstone documents for world-wide naval operations are the 2007 *A Cooperative Strategy for 21st Century Seapower (CS-21)* and the *Naval Operations Concept 2010 (NOC 10)*. These documents were jointly created between the Navy, Marine Corps, and Coast Guard and describe "when, where and how U.S. naval forces will contribute to enhancing security, preventing conflict and prevailing in war." If *CS-*

---

[13] North American Aerospace Defense Command and United States Northern Command, Arctic Collaborative Workshop: Arctic Oil Spill & Mass Rescue Operation - Tabletop Exercise, After Action Report (Colorado Springs, CO: USNORTHCOM, 2012), p. 29.

[14] Charles Emmerson, *The Future History of the Arctic* (New York: Public Affairs, 2010), p. 113.

21 describes the maritime strategy "end states" for naval forces, then *NOC-10* describes the "ways."[15] Taking its guidance from higher strategic documents, these two documents explain how the prosperity of the U.S. is inescapably tied to the security of the world's oceans. *NOC-10* goes on to explain the vital nature that security will play in the Arctic and[16] "supports mechanisms that underpin maritime security . . . and international law including the U.N. Convention on the Law of the Sea."[17] As evidenced in this document, service specific strategic documents are becoming more vocal in overt support of U.S. accession into UNCLOS.

The Navy was already looking at the implications of an ice free Arctic as early as 2000. In 2009, the Vice Chief of Naval Operations formed the Task Force Climate Change (TFCC). The specified tasks of the group's charter were to:

> recommend policy, strategy, roadmaps, force structure, and investments for the Navy regarding the Arctic and Climate Change that are consistent with existing National, Joint, and Naval guidance, including National Security Presidential Directive/Homeland Security Presidential Directive (NSPD-66/HSPD-25) . . . the initial focus for TFCC will be the Arctic, and the primary deliverable will be a holistic, chronological roadmap for future Navy action with respect to the Arctic between now and 2040.[18]

Informed by *NSPD-66/HSPD-25*, the TFCC promulgated the *Navy Arctic Roadmap* in 2009. Just like *NSPD-66/HSPD-25,* the *Navy Arctic Roadmap* identifies the Arctic as primarily a maritime domain. This cornerstone naval document sets ambitious timelines for the Navy to define objectives, assess operational capabilities, create strategy, and plan for future budgetary cycles. The document is intended to stay in effect until DOD

---

[15] Departments of Defense and Homeland Security, U.S. Navy/U.S. Marine Corps, U.S. Coast Guard, *Naval Operations Concept*, Department of Defense, (Washington DC, 2010), p. 1.
[16] Ibid., p. 32.
[17] Ibid., p. 37.
[18] Department of Defense, U.S. Navy, *Task Force Climate Change Charter*, October 30, 2009, (Washington DC: Department of Defense), p. 2.

completes the 2014 *Quadrennial Defense Review* (QDR) and will be revised after the promulgation of every subsequent QDR. While not openly advocating for U.S. accession into UNCLOS, the Navy recognizes the potential benefits of the U.S. being a ratified member to UNCLOS. The *Navy Arctic Roadmap* recognizes that a "changing environment and competition for resources may contribute to increasing tension . . . . Therefore, this roadmap considers the requirement for the governance framework provided by the United Nations Convention on the Law of the Sea (UNCLOS)."[19]

One of the action items tasks the TFCC to "provide support for U.S. accession to UNCLOS as applicable to Navy's interests in the Arctic." These tasks are to include "expression of Navy interest in the areas for which UNCLOS provides effective governance: freedom of navigation, treaty vs. customary law, environmental laws, and extended continental shelf claims. Development of talking points, information papers, or briefings for senior Navy leadership and Congressional staffs as requested."[20]

In 2010 the Navy published the *Navy Strategic Objectives for the Arctic* as the first deliverable directed by the *Navy Arctic Roadmap*. This guiding document states that the "Navy's desired end state is a safe, stable and secure Arctic region where U.S. national and maritime interests are safeguarded and the homeland is protected."[21] Security is the primary building block for U.S. Arctic policy and is a recurring theme that can be nested into higher strategic guidance such as the *NSS* and *NSPD-66/HSPD-25*. To achieve this end-state, the document outlines five strategic objectives and the desired effects required to achieve them.

---

[19] Department of Defense, U.S Navy, *Navy Arctic Roadmap*, (Washington DC: Department of Defense, October 2009), 6.
[20] Ibid., 11.
[21] Department of Defense, U.S. Navy, *Navy Strategic Objectives for the Arctic,* (Washington DC: Department of Defense, 21 May 2010), p. 1.

1. Contribute to safety, stability, and security in the region.
2. Safeguard U.S. maritime interests in the region.
3. Protect the American people, our critical infrastructure, and key resources.
4. Strengthen existing and foster new cooperative relationships in the region.
5. Ensure Navy forces are capable and ready.[22]

Two of the objectives listed in the *Navy Strategic Objectives for the Arctic* are of special note. The second strategic objective of safeguarding U.S. maritime interests in the region states that UNCLOS contains the appropriate framework for allowing the U.S. to successfully operate in the Arctic. It goes on to state that "U.S. accession to UNCLOS will enable and enhance the Navy's ability to protect our interests worldwide."[23]

The fifth strategic objective of ensuring Navy forces are capable and ready acknowledges that the Navy's missions continue to grow while allocation of limited resources remains a concern. The Navy understands that while they have some experience at operating in the Arctic, "the lack of environmental awareness, navigational capabilities, and supporting infrastructure, as well as competing jurisdictional and resource claims, are significant challenges that must be overcome by naval forces."[24] The Navy has submarines and aircraft that can patrol the Arctic on a needed basis, but has rightly identified that it lacks surface assets that could provide a constant presence except for select areas during ice free periods; however, no solutions have been identified to fill this possible requirement. The long awaited 2014 *Capabilities Based Assessment (CBA)* may shed light on how or even if the Navy plans to address this issue.

---

[22] Ibid., 2.
[23] Ibid., 3.
[24] Departments of Defense and Homeland Security, U.S. Navy/U.S. Marine Corps, U.S. Coast Guard, *Naval Operations Concept*, Department of Defense, (Washington DC, 2010), p. 32.

The U.S. Coast Guard

The U.S. Coast Guard has a long history of operating in the Arctic. Most of the increasing human activity in the Arctic Ocean basin points directly toward the missions of the U.S. Coast Guard. As one retired Coast Guard Admiral stated, "The smallest of the U.S. armed services shoulders the burden for a preponderance of the nation's maritime affairs, and the issues emerging in today's Arctic fall squarely upon the Coast Guard."[25]

One of the Coast Guard's eleven statutory missions deals with ice and the Polar Regions. In a combination of authorities from United States Code (USC) 14, 15, and 16, "the Coast Guard shall develop, maintain, and operate with due regard to the requirements of national defense, aids to navigation, icebreaking facilities, and rescue facilities for the promotion of safety on and over the high seas and waters subject to the jurisdiction of the United States."[26] As human activity increases in the Arctic, the Coast Guard will be required to conduct increased environmental protection, maritime domain awareness, and search & rescue in the Arctic. All of these mission sets are highlighted as areas of U.S. Arctic focus in *NSPD-66/HSPD-25*.

During a 64-day voyage in 1957, *USCGC SPAR* and *USCGC STORIS* were the first American vessels to cross the Arctic from the Pacific to the Atlantic via the Northwest Passage. In 1965, the Coast Guard was tasked with developing ice breakers for use in the Arctic and in Antarctica. For this purpose, all Navy ice breakers were transferred to the Coast Guard. Two years later, *USCGC POLAR STAR* and *USCGC*

---

[25] Rear Admiral Jeffrey M. Garrett (retired), "Enduring Arctic Reluctance", *sldinfo.com*, July 07, 2011, http://www.sldinfo.com/ending-reluctance/ (accessed November 12, 2012).
[26] United States Congress, *United States Code Title 14 - Coast Guard*, Vol. chapter 393, 1, 63 statute 495 (Washington, DC: August 4, 1949), p. 2.

POLAR SEA were commissioned. President G. W. Bush pressed Congress to fund an additional icebreaker.[27] In 1999, the Coast Guard commissioned its third icebreaker, USCGC HEALY. In 1994, USCGC POLAR SEA, with assistance from the Canadian Coast Guard, became the first U.S. surface vessel to reach the North Pole.

The Coast Guard, similar to the Navy, has begun to study the changing physical and strategic environment in the Arctic. In the *United States Coast Guard 2012 Posture Statement*, the Commandant of the Coast Guard, Admiral Robert J. Papp, outlined the Coast Guard's plans to "forward deploy to assist in maintaining the safety and security of anticipated Arctic exploratory oil drilling activity. In fiscal year 2013, the Coast Guard will begin the acquisition of a new Polar Icebreaker and invest in Alaskan infrastructure."[28] The Coast Guard commissioned a study entitled *United States Coast Guard High Latitude Region Mission Analysis Capstone Summary,* better known as the *High Latitude Study*. The purpose of this document was to "inform key decision makers evaluating upcoming acquisition and sustainment decisions for the Coast Guard's fleet of icebreaking vessels and associated aircraft, communications and forward operating locations."[29] At present, the U.S. has only three government owned heavy or medium ice breaking hulls. Of those three, two are past their 30-year life expectancy (*USCGC POLAR SEA* and *USCGC POLAR STAR*) and are currently non-operational. The third, *USCGC HEALY*, is a medium ice breaker with limited use in heavy ice conditions. There are currently several stop-gap measures being considered to keep at least one of the two

---

[27] George H. W. Bush, "Letter to Congressional Leaders Transmitting a Report on Polar Icebreaker Requirements December 21, 1990," American Reference Library - Primary Source Documents (01, 2001), p. 1.

[28] U.S. Department of Homeland Security, U.S. Coast Guard, *United States Coast Guard 2012 Posture Statement*, February 09, 2012, (Washington, DC: Department of Homeland Security), p. 29.

[29] ABS Consulting, United States Coast Guard High Latitude Regions Mission Analysis Capstone Summary, (Arlington, VA: ABS Consulting, 2010), p. 1.

POLAR class ice breakers operational, but those initiatives are costly and do not address the operational needs of the future. In the fiscal year (FY) 2013 budget, Admiral Papp has requested $6.1M for shore infrastructure in Alaska and $8.0M to initiate acquisition of the next generation of ice breakers.[30] The final analysis of the High Latitude Study found that the Coast Guard needed a total of six ice breakers, three heavy and three medium, to fulfill its statutory missions. This number would satisfy Arctic winter and transition season demands and provide sufficient capacity to also execute summer missions.

As a means to evaluate the mission analysis and perceived capabilities gaps found in the *High Latitude Study*, the Coast Guard created and executed Operation ARCTIC SHIELD 2012. This four month operation ending in October 2012 allowed the Coast Guard to focus on operations, community outreach to Alaskan Native partners, and an assessment of capabilities for operating in the Arctic. With the genesis coming from Royal Dutch Shell and their plans to begin exploratory drilling in September 2012, the Coast Guard developed a robust maritime safety and security plan to handle possible problems stemming from the drilling. This provided an excellent opportunity to test nearly every type of asset the Coast Guard has in its inventory; from cutters and boats, to aircraft, to environmental pollution response equipment. To support future Arctic operations, the Coast Guard established a seasonal helicopter base in Barrow consisting of two MH-60 Jayhawk helicopters and crews in 2012. Hailed as a logistical and operational success, the Coast Guard Seventeenth District Commander, Rear Admiral Tom Ostebo, remarked "for the first time, we had Coast Guard crews standing the watch

---

[30] U.S. Department of Homeland Security, U.S. Coast Guard, *United States Coast Guard 2012 Posture Statement*, February 09, 2012, (Washington, DC: Department of Homeland Security), p. 34.

and ready to support search and rescue, environmental protection and law enforcement operations in the Arctic."[31] In preparation for next summer, the Coast Guard is making plans to conduct Operation ARCTIC SHIELD 2013.

The United States currently lacks sufficient surface and air assets, maritime domain awareness, command and control structure, or Arctic infrastructure to meet the goals of *NSPD-66/HSPD-25*. The *Navy Operations Concept* states that icebreakers establish presence in international waters and "are the only means of providing assured surface access in support of Arctic maritime security."[32] The U.S. Navy possesses no icebreakers or ice hardened hulls that can operate in any type of ice.

In 2011, the Congressional Research Service (CRS) authored an issue paper on ice breaker modernization. The paper identified five missions that Coast Guard ice breakers can support in the Arctic. These five missions clearly mirror strategic guidance from the *National Security Strategy* and *NSPD-66/HSPD-25*. The five missions are:[33]

1. conducting and supporting scientific research in the Arctic and Antarctic;
2. defending U.S. sovereignty in the Arctic by helping to maintain a U.S. presence
3. in U.S. territorial waters in the region;
4. defending other U.S. interests in Polar Regions, including economic interests in
5. waters that are within the U.S. exclusive economic zone (EEZ) north of Alaska;
6. monitoring sea traffic in the Arctic, including ships bound for the United States;
7. And conducting other typical Coast Guard missions (such as search and rescue, law enforcement and protection of marine resources) in Arctic waters, including U.S. territorial waters north of Alaska.

---

[31] Department of Homeland Security, United States Coast Guard, "Coast Guard Completes Arctic Shield 2012," *uscgnews.com*, http://www.uscgnews.com/go/doc/4007/1594651/Imagery-Available-Coast-Guard-completes-Arctic-Shield-2012 (accessed November 11, 2012).
[32] ABS Consulting, United States Coast Guard High Latitude Regions Mission Analysis Capstone Summary, (Arlington, VA: ABS Consulting, 2010), p. 12.
[33] Congressional Research Service, *Coast Guard Polar Icebreaker Modernization Background, Issues, and Options for Congress* (Washington, D.C.: Government Printing Office, 2008), p. 3.

In 2008, the cost estimate for building a new ice breaker was between $800M and $925M and would take as long as eight to ten years to be built. In contrast, the estimate to extend the lives of the existing heavy ice breakers is approximately $400M per vessel.[34] Rear Admiral Jeffrey Garrett, U.S. Coast Guard ret., has argued for the recapitalization of the ice breaker fleet over building up conventional maritime forces in the Arctic; calling ice breakers a "Swiss army knife" capability. He went on to argue that:

> An on-scene icebreaker can respond to emergencies at sea and ashore, provide regulatory oversight of energy activities, monitor vessels near U.S. waters, facilitate scientific research, oversee and support oil spill responses, and exercise contingency plans with defense, local and international partners. In short, a patrolling icebreaker would assert a visible and capable U.S. presence. Augmenting the ship with special teams or additional personnel would tailor the response specific situations demand. An on-scene icebreaker would address every one of the *NSPD-66/HSPD-25* implementation policies.[35]

The reality of an ice free Arctic is rapidly approaching. Even with minimal or no ice, the Arctic is a dangerous environment to operate in. As shipping lanes become more open and countries begin to operate in the Arctic, the United States will not have the capabilities to provide presence necessary to enforce sovereignty issues. The rest of the Arctic world has long been preparing for this contingency. Russia, Canada, the Scandinavian countries, and even China have been working on building ice breaker fleets that are capable of pursuing their national interests in the Arctic.

The U.S. faces several strategic problems in the Arctic with respect to policy, capabilities, and command and control. First as previously stated, U.S. Arctic policy is

---

[34] Ibid., 12.
[35] Rear Admiral Jeffrey M. Garrett (retired), "Enduring Arctic Reluctance", *sldinfo.com*, July 07, 2011, http://www.sldinfo.com/ending-reluctance/ (accessed November 12, 2012).

vague. In an attempt to create a new national Arctic policy, *NSPD-66/HSPD-25* does little more than identify the importance of the Arctic and apply the end states from the *National Security Strategy* with an Arctic flavor. The policy lacks any concrete framework that could be built upon to accomplish those end states. Secondly, even though USNORTCHOM has been assigned as the principle military advocate for the Arctic, the command and control relationships are still unclear. Because of overlapping authorities, there is no single organization or chain of command for an Arctic contingency. Third, U.S. civilian and military entities have conducted dozens of Arctic studies and assessments. The overwhelming conclusion of these studies is that the U.S. is unprepared to conduct operations in the Arctic. The two major organizations that would operate in the Arctic, the Navy and the Coast Guard, have very little capability to meet national policy end states such as projecting a military presence, economic exclusion zone enforcement, or providing maritime environmental stewardship.

The U.S. will continue to struggle in identifying and protecting its interests in the Arctic without a comprehensive national strategy. The current approach of relying on customary law in the Arctic makes it difficult to establish any type of workable policy. The UNCLOS framework provides several strategic benefits to the United States. First and perhaps most importantly, it provides the foundation for defining U.S. sovereignty in the Arctic. Defining sovereignty is the vital first step in satisfying U.S. commerce concerns in the Arctic as it establishes the highly desired legal certainty necessary for U.S. businesses to operate in the region. Concurrently, answering basic sovereignty could be the catalyst for the creation of a more concrete national policy. U.S. stakeholders (military, law enforcement, business) could then begin developing ways and identifying means that would be informed by that more stable and detailed national policy.

CHAPTER 4: UNITED NATIONS CONVENTION ON THE LAW OF THE SEA

The United Nations Convention on the Law of the Sea (UNCLOS) was created to codify various customary maritime laws and establish responsibilities and rights of nations. As of 2013, the convention has been ratified by over 160 countries. The United States currently is not a ratified member of UNCLOS.

The history of customary maritime law goes back to the seventeenth century and the principle of the freedom of the seas. Commonly referred to as the "cannon shot" rule because of the approximate distance a shot could be fired from the shore, coastal nations informally claimed a three nautical mile sovereign buffer from the coastline outward to sea. Beyond that, the seas were considered free to all nations to exploit for commerce and subsistence.

In 1945, President Truman authored *Proclamation 2667*: *Policy of the United States with Respect to the Natural Resources of the Subsoil and Sea Bed of the Continental Shelf identifies*. By 1945, advances in technology allowed for profitable oil production in deeper waters further from shore. The U.S. recognized this and unilaterally decreed that the "Government of the United States regards the natural resources of the subsoil and sea bed of the continental shelf beneath the high seas but contiguous to the coasts of the United States, subject to jurisdiction and control."[1] Concurrently, President Truman issued *Proclamation 2668: Policy of the United States with Respect to Coastal Fisheries in Certain Areas of the High Seas*. Similar to Proclamation 2667, this document identified the need to protect resources and unilaterally claimed sovereign U.S.

---

[1] U.S. President. Proclamation, "Proclamation 2667: Policy of the United States with Respect to the Natural Resources of the Subsoil and Sea Bed of the Continental Shelf," *Federal Register* 10, no. 12305 (September 28, 1945).

rights to fishing grounds outside of three nautical miles.[2] These two documents together became known as the "Truman Proclamations." At the same time, other countries were beginning to make various claims about maritime sovereignty in order to gain access to more resources. With no common framework for defining equitable maritime sovereignty, some states believed that the more powerful countries were attempting to monopolize natural resources. In an effort to create a common framework for the international community, the United Nations formed the first United Nations Convention of the Sea (UNCLOS I) and produced four treaties. These four documents became known as the 1958 Geneva Conventions.

1. The Convention on the Territorial Sea and Contiguous Zone: Established a states right to a sovereign territorial sea; however, the treaty did not set a standard on how far that from shore that territorial sea was.

2. The Convention on the High Seas: Discussed freedom of navigation outside of territorial seas.

3. The Convention on Fishing and Conservation of the Living Resources of the High Seas: Discussed the right of states to fish in waters outside of other states continental shelf and laid out procedures on how to negotiate with a state to fish inside their sovereign continental shelf.

4. The Convention on the Continental Shelf: Was the first document to establish sovereign rights to resources outside of territorial seas.

UNCLOS I was generally seen as a success as it was the first international agreement to incorporate centuries of various customary maritime laws into codified international law. After what was largely seen as an unproductive failure at UNCLOS II in 1960, the United Nations once again convened to discuss maritime issues as part of UNCLOS III in 1967. UNCLOS III is the most current iteration and is the foundation for modern international maritime law. Officially starting in 1973, this nine-year conference

---

[2] U.S. President. Proclamation, "Proclamation 2668: Policy of the United States with Respect to Coastal Fisheries in Certain Areas of the High Seas," *Federal Register* 10, no. 12304 (September 28, 1945).

formally went into effect in 1994 after the required sixtieth signature, Guyana, ratified UNCLOS. One of the major advancements of UNCLOS III was that it improved upon the ideas set forth in the four treaties that made up the "1958 Geneva Conventions" and incorporated them all into UNCLOS III.[3] Three important parts of UNCLOS that relate to the Arctic include: the U.N. definition of freedom of navigation, the U.N. definition of an economic exclusion zone (EEZ), and the U.N. definition of the continental shelf.

UNCLOS Parts and Articles

Freedom of Navigation

Resolving one of the most ambiguous issues that previous iterations could not, UNCLOS III defined what area a state could claim as sovereign territorial sea. Part II article 3 of UNCLOS states that "every State has the right to establish the breadth of its territorial sea up to a limit not exceeding 12 nautical miles."[4]

The treaty recognized the need to create a standard that balanced a states sovereign rights and a states right to freedom of navigation. To that end, UNCLOS recognized Innocent Passage in Part II article 17. This allows foreign ships to conduct transit for expedience purposes through another state's territorial sea, "so long as it is not prejudicial to the peace, good order or security of the coastal State."[5] Under this article, foreign vessels can transit through other states 12 nautical mile territorial seas, but are prohibited (including military vessels) from conducting exercises, launching boats or

---

[3] United Nations Environment Programme (UNEP), "Background to UNCLOS," UNEP/GRID-Arendal, http://continentalshelf.org/about/1143.aspx (accessed November 28, 2012). The four treaties known as the "1958 Geneva Conventions" were formally nullified but the issues that they addressed were incorporated into UNCLOS III.

[4] United Nations, "United Nations Convention on the Law of the Sea," United Nations, http://www.un.org/Depts/los/convention_agreements/texts/unclos/closindx.htm (accessed August 26, 2012), Part II article 3.

[5] Ibid., Part II article 19.

aircraft, or carrying out research while transiting. Military ships are unable to deploy aircraft and submarines must transit on the surface to meet the requirement of Innocent Passage. Closely related to Innocent Passage is another provision called Transit Passage. Transit Passage in Part III article 37 allows a vessel to pass through an international strait that cannot be avoided for the purpose of expeditious transit. For instance, transiting ships are protected from impediment under Transit Passage to enter the Mediterranean Sea through the Strait of Gibraltar as it is considered an international strait. Unlike Innocent Passage, Transit Passage allows for military vessels to deploy aircraft and keep submarines submerged for self-defense purposes. Innocent and Transit Passage will become important principles in the Arctic as well. The Northwest Passage over Canada and the Northeast Passage (Northern Route) over Russia are rapidly becoming viable navigation routes during low ice and ice free periods. International discussions are already taking place on how to define these waterways. UNCLOS provides the necessary mechanisms to labels these potentially important sea lines of communication as internal waters, territorial seas, or international straits.

Economic Exclusion Zone

As fishing, oil production, and transportation technology grew in the twentieth century, states began to look further offshore. The potential of offshore oil resources became a catalyst to create a common framework for maritime sovereignty. Expanding on the principles created in the Convention on the Continental Shelf (one of the four treaties known collectively as the "1958 Geneva Conventions" that made up UNCLOS I), UNCLOS III clarified and defined a states sovereign rights over its continental shelf resources. That previously poorly defined area became known as an economic exclusion zone (EEZ). The newly created EEZ, as stated in Part V articles 56 and 57, grants coastal states "sovereign rights for the purpose of exploring and exploiting, conserving and managing the natural resources . . . of the waters superjacent [lying immediately above or upon] to the seabed and of the seabed and its subsoil. The exclusive economic zone shall not extend beyond 200 nautical miles from the baselines from which the breadth of the territorial sea is measured."[6] A coastal state would have complete sovereignty over all of the natural resources within its EEZ.

The major EEZs the U.S. could claim sovereignty over under UNCLOS included rich fishing stocks around New England and Alaska, as well as EEZs around island commonwealths and territories such as Puerto Rico and Guam. With new maritime sovereignty came the need for enforcement. Over the past thirty years, enforcement agencies, such as the U.S. Coast Guard and National Marine Fisheries Service, were needed to chase off or detain Russian ships fishing on the U.S. side of the Bering Strait and Gulf of Alaska. Although largely seen as an international success towards ensuring a coastal state's economic sovereignty over its maritime resources, the wording in UNCLOS is somewhat ambiguous when interpreting how to measure a states EEZ boundaries. This has been a point of contention between states with adjacent maritime borders.

As an example, the U.S. and Canada both claim a maritime area as part of their respective EEZ in the Arctic's Beaufort Sea. The disputed area is approximately 21,000 square kilometers, or about the size of Lake Ontario. The dispute arises from an overlap caused by the way each state interprets UNCLOS' method of defining an EEZ boundary. The dispute was mostly academic until recent years. Now that the Beaufort Sea is

---

[6] Ibid., Part V article 57.

experiencing ice free summers in certain areas, both countries are looking at the area for resource harvesting. The U.S. and Canada are strong allies and little danger of real conflict exists; however, there are several other international EEZ disputes involving other Arctic states, including Russia, which may be more difficult to resolve. Like the U.S – Canada dispute, these other disputes arise from overlap in individual claims.

Continental Shelf

UNCLOS III also defined and delineated the procedures for a coastal state to assert sovereignty over its continental shelf. Coastal states may claim out to 200 nautical miles nautical miles for its EEZ. However, the same state may be able to claim an additional 150 nautical miles (for a total of 350 nautical miles from shore) if it has an extended continental shelf (ECS) that meets the requirements in UNCLOS. Part VI article 76 (and follow on articles) of UNCLOS delineates two methods for defining a state's ECS. The two methods (or any combination of the two) used to define the ECS "shall not exceed 350 nautical miles from the baseline . . . or shall not exceed 100 nautical miles from the 2,500 meter isobaths (depth of 2,500 meters)."[7] The article defines the continental shelf as an area that "comprises the seabed and subsoil of the submarine areas that extend beyond its territorial sea throughout the natural prolongation of its land territory to the outer edge of the continental margin."[8] The coastal state proving a legitimate ECS has sovereignty over the natural resources listed in Part VI article 77 that consist of "mineral and other non-living resources of the seabed and subsoil together with living organisms belonging to sedentary species . . . . organisms which . . . are immobile on or under the seabed or are unable to move except in constant

---

[7] Ibid., Part VI article 76.
[8] Ibid., Part VI article 76.

physical contact with the seabed or the subsoil."[9] A state's extended continental shelf and economic exclusion zone are similar but are two separate and distinguishable maritime areas. With the exception of exclusive fishing rights, the rights granted to a state over the natural resources in its ECS are the same as the rights granted to a state in its EEZ. Approximately 80 coastal states have the naturally sloping seabed needed to meet UNCLOS requirement to claim an ECS.

Perhaps the most internationally and economically important issue in Part VI of UNCLOS deals with the subject of offshore drilling. Per Part VI article 81, "the coastal State shall have the exclusive right to authorize and regulate drilling on the continental shelf for all purposes."[10] One of the key items in Part VI on UNCLOS is article 82. This article requires that a country that exploits resources past the 200 nautical mile EEZ line will be required to share a percentage of those profits with the U.N., which in turn will be spread to other countries. This is another difference between an EEZ and an ECS. A state is not required to cede any profit from natural resources harvested in its EEZ. The profits will be ceded to the U.N. who will "distribute them to States parties to this Convention, on the basis of equitable sharing criteria, taking into account the interests and needs of developing States, particularly the least developed and the land-locked among them."[11]

UNCLOS Governing Entities

Several international bodies were created under the UNCLOS framework that were designed to promote economic equity among states, defuse conflict, and facilitate resolution of disputes between states. Most notably the Commission on the Limits of the

---

[9] Ibid., Part VI article 77.
[10] Ibid., Part VI article 81.
[11] Ibid., Part VI article 82.

Continental Shelf (CLCS), the International Tribunal for the Law of the Sea (ITLOS) and other dispute mechanisms, and the International Seabed Authority (ISA). UNCLOS and ITLOS provide states an outlet to discuss grievances in a peaceful manner using a common and agreed upon framework.

Commission on the Limits on the Continental Shelf (CLCS)

Drawing its mandate from Annex III article 2 of UNCLOS, the Commission on the Limits of the Continental Shelf (CLCS) is the body that considers the scientific data submitted by coastal states concerning the boundaries of its extended continental shelf (ECS) in areas where it extends beyond 200 nautical miles (outside of the EEZ).[12] The CLCS is a scientific body that assists in validating claims of maritime sovereignty and the U.N. maintains that the commission is "not a court of law, nor was it ever expected to be one . . . . The role of this highly scientific organ, which is called upon to provide assistance in the much politicized realm of setting legal boundaries, is to help establish the true limit of the outer boundary of the continental shelf according to the terms of UNCLOS."[13] To date, there have been 63 ECS submissions from the international community to the CLCS. Six of those claims have been from Arctic nations.[14]

To make a claim to the CLCS, a state must be a ratified member of UNCLOS. Not being a ratified party to UNCLOS, the United States is prevented from submitting an ECS claim to the CLCS, and conversely, is unable to provide input on the legitimacy on

---

[12] United Nations Division for Ocean Affairs and the Law of the Sea, "Commission on the Limits of the Continental Shelf," United Nations, http://www.un.org/Depts/los/clcs_new/clcs_home.htm (accessed October 19, 2012).

[13] Myron H. Nordquist and University of Virginia. Center for Oceans Law and Policy. Conference, "The Law of the Sea Convention : US Accession and Globalization" (Boston: Martinus Nijhoff Publishers, 2012), p. 225.

[14] United Nations, "Submissions to the Commission on the Limits of the Continental Shelf", *www.un.org*, http://www.un.org/Depts/los/clcs_new/commission_submissions.htm: (accessed October 19, 2012).

other states submissions. This will prove problematic for the United States in garnering international recognition for its Arctic ECS. The U.S. may have more to gain than any other Arctic state. With naturally shallow depths on the Alaska North Slope, "our continental shelf could extend 600 miles into the Arctic."[15]

International Tribunal on the Law of the Sea (ITLOS) and other Dispute Resolution Mechanisms

UNCLOS encourages states to resolve their disputes peacefully through bilateral means. Recognizing that this is not always achievable, UNCLOS established mechanisms for resolving disputes between states on maritime issues. Part XV article 287 outlines the four different venues for peaceful resolution. These include the International Tribunal for the Law of the Sea (ITLOS), the International Court of Justice, arbitration, and special arbitration.

The International Tribunal for the Law of the Sea serves as a forum to resolve disputes between states over application or interpretation of UNCLOS articles. This body hears cases dealing with a wide range of issues such as fishing rights, international pollution, and freedom of navigation. Many of these cases involve the alleged violation of resource harvesting within a state's EEZ or territorial seas. The creation of ITLOS is a new phenomenon to the international maritime community. Previous to the creation of ITLOS, states had no other venue to address grievances except through direct contact with another state. In some cases, this had led to conflict.

---

[15] Senate Foreign Relations Committee, "Accession to the 1982 Law of the Sea Convention and Ratification of the 1994 Agreement Amending Part XI of the Law of the Sea Convention," http://www.foreign.senate.gov/imo/media/doc/REVISED_Secretary_Clinton_Testimony.pdf p.2 (Accessed March 10, 2013).

In 2010, Russia and Norway were able to resolve a long standing Arctic EEZ dispute using UNCLOS framework. An area in the Barents Sea, approximately half the size of Germany, has been an area of friction between the two nations since the 1970s. The dispute originally began over fishing rights but became even more contentious when oil was discovered. Both countries agreed to a treaty that split the area in half.

Since the U.S. is not a ratified party to UNCLOS, it does not have the luxury of utilizing three of the four dispute resolution mechanisms laid out in UNCLOS (the U.S. can still use the International Court of Justice because it is a member of the United Nations). The United States will be limited in its ability to conduct dispute resolution with other Arctic states as every other Arctic state (besides the U.S.) is party to UNCLOS. Because of this, the U.S. will have to rely almost solely on bilateral diplomacy in dealing with any future Arctic dispute.

International Seabed Authority

Part XI of UNCLOS deals with international waters that cannot be claimed as sovereign by any state. Largely in response to Part XI of UNCLOS, ITLOS created a separate chamber known as the International Seabed Authority (ISA), which became an autonomous entity in 1996. The ISA was created to provide administration for deep sea mining in international waters. Part XI of UNCLOS has been a contentious topic for the U.S. since its creation and the principal stated reason why the U.S. did not ratify UNCLOS at the close of the convention in 1982. Part XI specifically deals with the waters outside of any state's extended continental shelf or economic exclusion zone that cannot be claimed as sovereign by any state. As stated in Part XI article 136, the Area and its resources are the "common heritage of mankind."[16] The article goes on further to say that no state may make a claim of sovereignty to any part of the Area or its resources, and that any gains made in the Area must be made and shared with the international community.[17] Article 144 includes the provision that "the ISA shall take measures . . . promote and encourage the transfer to developing states technology and scientific knowledge so that all state parties benefit from them."[18] President Reagan did not agree with the provisions set forth in Part XI, most notably "in the deep seabed mining area, we will seek changes necessary to correct those unacceptable elements and to achieve the goal of a treaty that . . . will be likely to receive the advice and consent of the Senate. In this regard, the convention should not contain provisions for the mandatory transfer of private technology."[19] He argued that Part XI was not in the best interests of the U.S. or the free market system. President Reagan acknowledged that the United States supported UNCLOS as customary international law, with the exception of Part XI, and vowed to continue to work with the convention to produce a product that was acceptable to U.S. interests. A year later, President Reagan proclaimed that the U.S. remained opposed to the provisions in Part XI in UNCLOS but would "recognize the rights of other states in the waters off their coasts, as reflected in the Convention, so long as the rights and freedoms of the United States and others under international law are recognized by such coastal states."[20]

---

[16] United Nations, "United Nations Convention on the Law of the Sea," United Nations, http://www.un.org/Depts/los/convention_agreements/texts/unclos/closindx.htm (accessed August 26, 2012), Part XI article 136.
[17] Ibid., Part XI article 137.
[18] Ibid., Part XI article 144.
[19] U.S. President, Proclamation, "President Ronald Reagan's Statement on United States Participation in the Third United Nations Conference on the Law of the Sea," (January 29, 1982).
[20] U.S. President, Proclamation, "Statement on United States Oceans Policy," (March 10, 1983).

In an effort aimed at removing U.S. opposition to UNCLOS due to Part XI and paving the way for U.S. ratification, the convention negotiated the *Agreement Relating to the Implementation of Part XI of the Convention*. This separate document was a political victory for the U.S. as all of the "Reagan administration's objections were fixed to the satisfaction of the United States."[21] The U.S. would no longer be required to share technology with other states and it would gain a permanent seat on the ISA that included veto power over all ISA decisions (the only permanent seat on the ISA). President Bill Clinton signed the *Agreement Relating to the Implementation of Part XI of the Convention* in 1994 and sent both (the Agreement and UNCLOS) to the Senate. Neither document ever made it to the Senate floor for discussion.

---

[21] Scott G. Borgerson, "The National Interests and the Law of the Sea," *Council on Foreign Relations: Council Special Report no. 46*, (May 2009): p. 12.

CHAPTER 5: THE UNITED STATES AND UNCLOS

There are currently 164 states that are ratified members of UNCLOS. The United States is among a list of non-party states that includes Libya, Iran, and North Korea. The arguments in the battle for and against U.S. accession into UNCLOS are many and varied. Those who are against U.S. accession to UNCLOS are typically conservative Republicans who see this treaty as ceding sovereignty to the United Nations. In 2004, Jeanne Kirkpatrick, former Ambassador to the United Nations, testified before the Senate Armed Services Committee that "ratification [of UNCLOS] will diminish our capacity for self-government, including, ultimately, our capacity for self-defense."[1] The main tenet of this argument is that of the protection of the individual nation-state. Secretary of State George Shultz summed up the conservative argument against ceding sovereignty:

> The world has worked for three centuries with the sovereign state as the basic operating entity, presumably accountable to its citizens and responsible for their well-being. In this system, states also interact with each other to accomplish ends that transcend their borders. They create international organizations to serve their ends, not govern them.[2]

In 2007, President G.W. Bush parted ways with the conservative mainstream urging Congress to ratify UNCLOS:

> Joining will serve the national security interests of the United States, including the maritime mobility of our armed forces worldwide. It will secure U.S. sovereign rights over extensive marine areas, including the valuable natural resources they contain. And it will give the United States a seat at the table when the rights that are vital to our interests are debated and interpreted.[3]

---

[1] Senate Armed Services Committee, "Jeane J. Kirkpatrick, Testimony before the Senate Armed Services Committee" April 8, 2004.
[2] George P. Schultz, "A Changed World", lecture - Library of Congress, Washington, DC, March 22, 2004.
[3] President, Proclamation, "President's Statement on Advancing U.S. Interests in the World's Oceans" (Lanham, United States, Lanham: Federal Information & News Dispatch, Inc., 2007),

In summer of 2012, the Senate Foreign Relations Committee overwhelmingly voted to bring the vote for ratification to the Senate floor. After a group of 34 conservative Senators, led by Senator Jim DeMint of South Carolina, signed a letter saying that they would vote no on UNCLOS, the vote never materialized.[4]

Outside of sovereignty concerns, opponents argue that the treaty is unnecessary. The United States already has worldwide freedom of navigation and authority to exploit undersea resources under the blanket of customary international law.

The conservative think tank Heritage Foundation has been a vocal advocate for U.S. unilateralism in the Arctic, with the battle cry for defeating UNCLOS tending to start with statements such as "National sovereignty should be the cornerstone of U.S. Arctic policy. In the Arctic, sovereignty equals security and stability."[5] There is a concern that United States could be in danger of becoming subject to an international organization that will be making decisions for the majority of the world; those decisions may not be in the nation's best interest as "the U.N. majority does not have much use for us. That majority was responsible for creating the present Law of the Sea Treaty . . . but also has supra-national government, with an executive, with a legislature, with a judiciary."[6]

Spring Baker of the Heritage Foundation pointed out that the world has seen too many failures from the United Nations, such as the oil for food program and the

---

http://search.proquest.com.ezproxy6.ndu.edu/docview/190573909?accountid=12686 (accessed December 16, 2012).

[4] Zack Colman, "Republican Senator Says Sea Treaty might Pass After Election," *The Hill*, August 17, 1012.

[5] Luke Coffey, "Arctic Region: U.S.Policy on Arctic Security," *heritage.org*, http://www.heritage.org/research/reports/2012/08/arctic-security-five-principles-that-should-guide-us-policy?rel=Arctic (accessed September 13, 2012).

[6] Frank J. Gaffney Jr., "U.N.'s Larger Role in UNCLOS is Bad for American Interests." *Texas Review of Law & Politics* 12, no. 2 (2008): p. 474.

subjugation of African women by UN peacekeepers. In his view, UNCLOS was created from the same cloth. He goes on to argue that if the United States had helped to craft a treaty that was built on the importance of the navigation rights created from previous conventions "and not adopted these institutions that would challenge state sovereignty, that would be anti-free market, that would pursue an anti-American agenda . . . it would have sailed through the Senate."[7]

Another argument is that United States military will lose its autonomy under UNCLOS. The Navy's use of SONAR has been condemned by environmentalists who say it is detrimental to the health of acoustic using marine mammals, such as dolphins and whales. There are those who argue that the Tribunal of the Law of the Sea could ban the use of SONAR by military forces.[8] This example, among others, reflects a belief that the United Nations has become an organization, in the words of Senator Jim Inhofe, an "overzealous international organizations with anti-American biases that infringe upon American society."[9]

Other opponents point to the power of the International Seabed Authority. In Part XI, UNCLOS authorized the transfer of technology to developing countries. Doug Bandow, who served as a representative to the Law of the Sea Convention in 1994, wrote that UNCLOS is "intended to inaugurate large and sustained wealth transfers from the

---

[7] Baker Spring, "All Conservatives should Oppose UNCLOS," *Texas Review of Law & Politics* 12, no. 2 (Spring 2008), p. 456.
[8] Frank Gaffney Jr., "U.N.'s Larger Role in UNCLOS is Bad for American Interests", *Texas Review of Law & Politics* 12, no. 2 (Spring 2008), p. 471.
[9] Amy Payne, "Republican Opposition Downs U.N. Disability Treaty," *usatoday.com*, http://www.usatoday.com/story/news/politics/2012/12/04/disability-united-nations-treaty-senate-dole/1745679/ (accessed February 02, 2013).

industrialized states . . . . LOST [law of the sea treaty] remains captive to its collectivist and redistributionist origins."[10]

Under UNCLOS an international organization has the power to levy a fee or a tax on sovereign states. Part VI article 82 of UNCLOS refers to payments and contributions to the International Seabed Authority that would be required by coastal states as part of the harvesting of resources past the 200 nautical mile EEZ.[11] This article contains ambiguous language describing how these funds would be distributed to other non-coastal states. The concern is that "we would be giving to a U.N. organization the ability to raise its own revenues."[12]

Effects of climate change in the Arctic may be the catalyst needed for advocates of the treaty to make the final push for ratification. In 2007, James Baker III, who served in various cabinet positions in the Reagan and G.W. Bush administrations, declared that U.S. ratification of UNCLOS was necessary to "define maritime zones, preserve freedom of navigation, allocate resource rights, and establish certainty necessary for various businesses that depend on the sea."[13] During 2012 committee hearings on UNCLOS, Secretary of Defense Leon Panetta strongly urged the Senate to ratify UNCLOS immediately, stating that the growing security challenges the United States faces "are

---

[10] Doug Bandow, "The Law of the Sea Treaty: Inconsistent with American Interests." CATO Institute. http://www.cato.org/publications/congressional-testimony/law-sea-treaty-inconsistent-american-interests (accessed December 09, 2012).

[11] United Nations, "United Nations Convention on the Law of the Sea," United Nations, http://www.un.org/Depts/los/convention_agreements/texts/unclos/closindx.htm (accessed August 26, 2012), Part VI Article 82.

[12] Frank Gaffney Jr., "U.N.'s Larger Role in UNCLOS is Bad for American Interests", *Texas Review of Law & Politics* 12, no. 2 (Spring 2008).p. 473

[13] Henry Kissinger, George Shultz, James Baker III, Colin Powell, and Condolezza Rice, "Time to Join The Law of the Sea Treaty," *online.wsj.com,* May 20, 2012. http://online.wsj.com/article/SB10001424052702303674004577434770851478912.html (accessed December 12, 2012).

beyond the ability of any single nation to resolve alone."[14] During testimony to the Senate Foreign Relations Committee in June 2012, the Chief of Naval Operations, Admiral Jonathan Greenert, asserted that an assurance of codified international law and an institutionalized process for dispute resolution greatly enhances the ability of the United States to deter aggression, contain conflict, and win the nation's wars. UNCLOS guarantees freedom of navigation in areas of strategic interest. The freedom of navigation established in UNCLOS is essential in keeping important sea lines of communication open.[15] General Charles Jacoby, commander of USNORTHCOM, explained that the future of U.S. security would rely heavily on cooperative partnerships. "From an Arctic perspective, our accession to the convention is important to encouraging cooperative relationships among Arctic states . . . . Future defense and civil support scenarios in the Arctic maritime domain will require closely coordinated, multinational operations."[16]

The Commandant of the Coast Guard, Admiral Robert Papp, also testified before Congress in support of U.S. accession into UNCLOS. In addition to maritime security and freedom of navigation, ADM Papp conveyed the need for accession to set international provisions on law enforcement, especially drug smuggling. Arguing that a non-party status hurts U.S. efforts in gaining cooperation via bi-lateral agreements with its international partners, ADM Papp explained that the provisions embedded in UNCLOS would "cement a common cooperative framework, language, and operating

---

[14] Leon Panetta, "UNCLOS Accession would Strengthen US Global Position," *Hampton Roads International Security Quarterly* (Jul 1, 2012), p. 23.
[15] Jonathan Greenert, "UNCLOS and U.S. Freedom of Navigation," *Hampton Roads International Security Quarterly* (Jul 1, 2012), p. 37.
[16] Charles H. Jacoby jr., "UNCLOS Vital to Arctic Cooperative Security," *Hampton Roads International Security Quarterly* (Jul 1, 2012), p. 45.

framework . . . in securing expeditious boarding, search enforcement, and disposition decisions" of law enforcement cases.[17] ADM Papp stated his belief that accession into UNCLOS was essential for the U.S. to pursue an Arctic strategy. He described UNCLOS as the "umbrella" necessary to the Coast Guard's statutory missions of environmental protection, maritime security, and law enforcement in the Arctic. Accession into UNCLOS, he said, "provides the legal framework we need to take advantage of opportunities."[18]

Speaking before the Senate Foreign Relations Committee in 2012, Thomas J. Donohue, the President and CEO of the U.S. Chamber of Commerce, testified that "joining the convention will provide the U.S. a critical voice on maritime issues from mineral claims in the Arctic to how International Seabed Authority funds are distributed . . . . Contrary to some opponents claims, joining the Treaty promotes American sovereignty. LOS strengthens our sovereignty by codifying our property claims in the Arctic and on our ECS [extended continental shelf]."[19] The business community claims that technology is at the point where it is financially feasible to exploit these resources; however "companies need the certainty the Convention provides in order to explore beyond 200 miles and to place experts on international bodies that will delineate claims in the Arctic."[20] The Chairman and CEO of Exxon, R. W. Tillerson, in a 2012 letter to

---

[17] Robert Papp, "Enhancing Coast Guard Operations through UNCLOS Accession," *Hampton Roads International Security Quarterly* (Jul 1, 2012), p. 47.
[18] Ibid.
[19] Thomas J. Donohue, "Before the Committee on Foreign Relations of the United States Senate - Testimony of Thomas J. Donohue, President and Chief Executive Officer, U.S. Chamber of Commerce", foreign.senate.gov, June 28, 2012, http://www.foreign.senate.gov/imo/media/doc/Donohue%20Testimony.pdf, p. 2, (accessed January 21, 2013).
[20] American Petroleum Institute, Chamber of Shipping of America, Financial Service Roundtable, International Association of Drilling Contractors, Marine Retailers Association of the Americas, National Association of Manufacturers, National Marine Manufacturers Association, North American Submarine

the Senate Foreign Service Committee, expressed his company's support for the ratification of UNCLOS as a necessity to financially and efficiently operate in the Arctic. He elaborated that there are currently overlapping claims in the Arctic and that UNCLOS provides the legal basis necessary for resolving claims and establishing stability necessary to support development. Otherwise, "the lack of legal certainty unnecessarily clouds our investment motivation."[21] Thomas J. Donahue of the U.S. Chamber of Commerce echoed Tillerson's statement in a January 2012 letter to Senators John Kerry and Richard Lugar, pointing out that without UNCLOS "no U.S. company will make the multi-billion dollar investments required to recover these resources without the legal certainty the Convention provides."[22]

In addition to exploiting the resources in a respective economic exclusion zone, Arctic countries are scrambling to map out their extended continental shelves. For the United States, this could produce billions, perhaps trillions, of dollars in profits from oil, natural gas, and minerals. Of great concern is the harvesting of seabed minerals in the form of rare earth metals: namely manganese, nickel, copper, and cobalt. In discussing rare earth metals and the need for ratification, the National Association of Manufacturers claims that "China produces more than 90 percent of the world's supply and also consumes roughly 60 percent . . . . China recently imposed significant export restrictions

---

Cable Association, RARE, The Association for Rare Earth, TechAmerica, Telecommunications Industry Association, U.S. Chamber of Commerce, Letter to Senators John Kerry and Richard Lugar supporting ratification of UNCLOS dated June 13, 2012, ratifythetreatynow.org. http://ratifythetreatynow.org/sites/default/files/pdf/Biz%20Support%20for%20Law%20of%20Sea%206-13-12.pdf (accessed January 21, 2013).

[21] Rex W. Tillerson, Letter to Senators John Kerry and Richard Lugar expressing support for UNCLOS dated June 08, 2012, *ratifythetreatynow.org*, http://ratifythetreatynow.org/sites/default/files/pdf/ExxonMobil%20%2806-08-12%29.PDF, (accessed January 21, 2013).

[22] Thomas J. Donohue, Letter to U.S.Senate expressing support for UNCLOS dated July 27, 2012, ratifythetreatynow.org, http://ratifythetreatynow.org/sites/default/files/pdf/Chamber%20of%20Commerce%20%2807-27-12%29.pdf, (accessed January 21, 2013).

on its rare earth production. In 2010, it announced it would cut exports by 40 percent in 2012."[23] These minerals are extremely important to the production of telecommunications, defense systems, and manufacturing. Without being a ratified member of UNCLOS, proponents of the treaty point out that the United States will not be heard in the policy making process. As a non-party the U.S. does not have a representative on the International Seabed Authority (ISA) or Commission on the Limits of the Continental Shelf (CLCS).

These same arguments extend to the exploitation of Arctic oil and natural gas as well. If the United States were to gain all of the undersea area that many believe it is entitled to through its extended continental shelf, that area could extend up to 600nm from the Alaska coastline. In addition, the U.S. could gain upwards of 4.1 million square miles of ocean floor, an area greater than the 48 contiguous states, the largest jurisdiction grant of any nation in the world.[24] Just the area within the EEZ around Alaska may hold as much as 27 billion barrels of oil and 132 trillion cubic feet of natural gas.[25] In total, the United States would have the largest EEZ/ECS area of any country in the world, one that extends into three separate oceans.

Shipping is another concern for UNCLOS advocates. With the opening of the Arctic, the international community is looking at the possibilities of shortened commerce transit routes that could save millions in time and money. Supporters argue that relying on existing customary maritime laws does not provide enough legal certainty for business

---

[23] Jay Timmons, "UNCLOS Critical for US Manufacturing Competitiveness," *Hampton Roads International Security Quarterly* (Jul 1, 2012), p. 103.

[24] Thomas J. Donohue, "Before the Committee on Foreign Relations of the United States Senate - Testimony of Thomas J. Donohue, President and Chief Executive Officer, U.S. Chamber of Commerce", *foreign.senate.gov*, June 28, 2012, http://www.foreign.senate.gov/imo/media/doc/Donohue%20Testimony.pdf, p. 3, (accessed January 21, 2013).

[25] Ibid.

to grow. Over 95 percent of U.S. commerce is transported by water. UNCLOS supporters point to the benefits of legally established territorial seas, the right of innocent passage, and unimpeded transit through archipelagos and international straits. Stable, long-term laws benefit business worldwide. In a letter to Secretary of State Hillary Clinton, the President of the Maritime Trades Department of the AFL-CIO, Michael Sacco, presented the argument that U.S. accession to UNCLOS creates a safer atmosphere for international shipping, as it "places an obligation on its signatories to do everything in their power to preserve high seas for innocent use."[26] Without ratifying UNCLOS, other countries could potentially have a voice in crafting international laws that are unfavorable to U.S. business.

UNCLOS has been a contentious issue in the U.S. since its creation. No one in the U.S. political arena appears to be wavering in their beliefs for or against UNCLOS; hence there are no signs of resolution on the horizon. America's wait-and-see approach to the Arctic continues to be at odds with the certainty of a changing Arctic environment. The U.S's inability to create concrete policy for the Arctic could eventually force the U.S. to make decisions on service capabilities that are undesirable. For instance, lack of policy may eliminate or defer acquisition projects for needed Arctic capabilities. What is not uncertain is that other countries are moving forward with succinct Arctic policies that will prepare them for upcoming military, economic, and political Arctic contingencies.

---

[26] Michael Sacco, Letter to Secretary of State Hillary Clinton dated September 21, 2011, *ratifythetreatynow.org*, http://ratifythetreatynow.org/sites/default/files/pdf/Letter-Sacco-September-2011.pdf, (Accessed January 21, 2013).

CHAPTER 6: U.S. STRATEGIC CONSIDERATIONS IN THE ARCTIC

Eight Arctic nations, all are party to the Convention, are actively working to secure their resource and sovereignty rights in the Arctic. Even though it is not an Arctic nation, China is moving forward with Arctic plans and capabilities. Each Arctic state views the region from a slightly different perspective, but they all share common basic concerns: sovereignty, security, economics, and stewardship.

Russia

Russia has always seen itself as inescapably tied to the Arctic. The region is considered the energy frontier of the Russian future. As much as 20 percent of Russia's gross domestic product (GDP) and 22 percent of total Russian energy exports are generated north of the Arctic Circle.[1] Holding over a quarter of the world's natural gas reserves, approximately 45 percent of those reserves are in the Siberian Arctic. With massive oil and natural gas reserves, Russia has overtaken Saudi Arabia as the largest oil producer, and has overtaken the United States as the largest natural gas producer.[2] The oil and gas industry is the strategic means for Russia to regain its former status as a world power. Even though Russia is a major developer of oil, natural gas is the prize that Russia eyes in the Arctic. It is estimated that there may be over 968 trillion cubic feet of natural gas in the Western Siberian and Eastern Barents Sea basins.[3] The economic potential for Russia is enormous; but, so is the cost for development of this resource. Natural gas is much more difficult to extract than crude oil and Gazprom does not

---

[1] Katarzyna Zysk, "Russia's Arctic Strategy," *JFQ: Joint Force Quarterly*, no. 57 (2010): 105.
[2] United States Energy Information Administration, "Russia," United States Department of Energy, http://www.eia.gov/countries/cab.cfm?fips=RS (Accessed December 22, 2012).
[3] Geology.com, "Oil and Natural Gas Resources of the Arctic," http://geology.com/articles/arctic-oil-and-gas/ (Accessed January 02, 2013).

possess the needed capital, infrastructure, or technology to operate in the Arctic profitably.

In 2007, a Russian expedition led by the nuclear powered icebreaker *RUSSIA* made its way to the North Pole and planted a Russian flag on the ocean floor in support of a claim of ownership in the region. While publicly dismissed by other Arctic nations as simple political theater, it was the first time that man had reached the ocean floor at the pole and the action resonated throughout the world. Russia justified the action by likening it to the American moon landing. In response, Canadian Prime Minister Peter MacKay said "this isn't the 15th century. You can't go around the world and just plant flags and say 'We're claiming this territory'."[4] Nevertheless, it was a clear signal from Russia of its intent and determination to move forward with an aggressive Arctic policy and that a "return to the Arctic" was a national priority.[5] Seeing the economic advantages emerging in the Arctic, Russia became party to the Convention in 1997. After a major scientific endeavor to map the ocean floor, Russia originally submitted its claim to the Commission on the Limits of the Continental Shelf (CLCS) in 2001. The undersea area of the Arctic that Russia has claimed is a massive area known as the Lomonosov Ridge. Claiming that the ridge is a natural extension of its continental shelf, this massive area would give Russian an additional 380,000 square miles of sea bed under its sovereignty. The CLCS returned the claim for additional scientific justification. Russia is expected to resubmit its updated claim in 2013.

---

[4] "Russia Plants Flag Under North Pole," *BBC News*, August 02, 2007, http://news.bbc.co.uk/2/hi/europe/6927395.stm, (Accessed January 12, 2013).
[5] Konstantin Voronov, "The Arctic Horizons of Russia's Strategy," *Russian Politics & Law 50, no. 2* (Mar, 2012): 56.

In the Russian assessment, there is no imminent threat of direct aggression against Russian territory from the Arctic or of a large-scale military confrontation in the region. Moscow does not rule out the possibility of competition for hydrocarbon reserves developing into tensions that may involve the use of military power.[6] To that end, Russia has resumed long range bomber flights near Scandinavia and Canada and an increased presence of the Northern Fleet in the Arctic.[7]

By far, Russia is the country most prepared to operate in the Arctic, with the largest percentage of their naval assets assigned to the Northern Fleet, including an aircraft carrier and nuclear submarines. Perhaps the most striking fact is the number and composition of Russia's icebreaker fleet. Russia currently possesses over 30 ice breakers of various sizes and strengths. Of those icebreakers, six are nuclear. Russia is the only nation in the world with nuclear icebreakers, which are used for scientific research and clearing shipping lines of communications. In August 2012, Russia signed a contract for a new 568 foot nuclear icebreaker, the largest in its fleet. It is expected to be completed in 2017.[8]

Russia is moving forward with a very nationalistic policy in the Arctic and could be a significant security and economic challenge for NATO, Europe, and the United States.

---

[6] Katarzyna Zysk, "Russia's Arctic Strategy," *JFQ: Joint Force Quarterly*, no. 57 (2010): 108.
[7] Michael L. Roi, "Russia: The Greatest Arctic Power?" *Journal of Slavic Military Studies* 23, no. 4 (Oct, 2010): 558.
[8] Katia Moskvitch, "Russia to Build Biggest Nuclear-Powered Icebreaker," *British Broadcasting Company*, http://www.bbc.com/news/technology-19576266 (Accessed September 17, 2012).

China

With no Arctic coast or sovereign claim to an Arctic EEZ, China currently has little influence on the policies being made by the Arctic Council and the United Nations. China applied for permanent observer status to the Arctic Council in 2009, and a vote is expected on that request in 2013. As a ratified member of UNCLOS, China's current position is that the Arctic should be available for all of mankind to use, despite the fact that China is in violation of UNCLOS over its sovereignty claims in the South and East China Sea. The same treaty that China is relying on to support ancestral claims of sovereignty in Asia may exclude it from access to resources in the Arctic. Said one high ranking Chinese Admiral, "the Arctic belongs to all the people around the world, as no nation has sovereignty over it . . . . China must play an indispensable role in Arctic exploration as we have one-fifth of the world's population."[9] Speaking to an Arctic forum in Europe in 2009, China's assistant minister of foreign affairs, Hu Zhenguye, publicly stated that "China does not have an Arctic strategy,"[10] yet China has been very active in Antarctica since the early 1980s and began active Arctic exploration, mostly for environmental studies, in 1995. The Chinese icebreaker *XUELONG*, the largest non-nuclear icebreaker in the world, conducted a historic exploration in 2010 and made it all the way to the 88th parallel and transported a team to the North Pole via helicopter for the first time. China has ordered an additional polar icebreaker that will be completed in 2013, making the Chinese icebreaker fleet larger than that of the United States.

---

[9] David Curtis Wright and Naval War College (U.S.). China Maritime Studies Institute., "*The Dragon Eyes the Top of the World Arctic Policy Debate and Discussion in China*," U.S. Naval War College, China Maritime Studies Institute p. 2

[10] Linda Jakobson and Stockholm International Peace Research Institute., "*China Prepares for an Ice-Free Arctic*," Stockholm International Peace Research Institute, p.9

As China's standard of living and the associated industrial demand continues to rise, along with its energy demands, the Chinese may, by necessity, be driven to the Arctic. While not overtly expressing a desire to begin exploration for resources in the Arctic, the Chinese are relentlessly searching for new partners in energy production in Africa, South America, and Australia. Many believe that if ready access to oil and natural gas become a reality in the Arctic, China will become much more vocal about being kept out. With a healthy distrust of Russia, China is beginning to look at Canada and Scandinavia for strategic partnerships.

Almost half of China's gross domestic product is dependent on the shipping industry.[11] With the possible opening of the Northeast Passage, China is very interested in using this route to shorten the transportation distance of Chinese exports and energy imports. In anticipation of creating a hub for international shipping, China began establishing friendly relations with Iceland in 2005. By 2007, China had the largest international embassy in Iceland and welcomed the President of Iceland to China with all the pomp normally reserved for a major head of state.[12] Some hypothesize that this goes beyond mere shipping interests, and that China is attempting to gain a strategic foothold near the Arctic. The Suez and Panama Canal both have restrictions on the size of ships that can transit through them. The Arctic has no such restrictions. With shorter transit times and the possibility of using larger ships, the financial gains of Arctic shipping for China could be substantial.

---

[11] Gao Weijie, "Development Strategy of Chinese Shipping Company Under the Multilateral Framework of WTO ," *cosco.com*, http://www.cosco.com/en/pic/forum/654923323232.pdf (Accessed February 02, 2013), p. 1.

[12] Robert Wade, "A Warmer Arctic Ocean Needs Shipping Rules," *Financial Times*, http://www.ft.com/intl/cms/s/0/1c415b68-c374-11dc-b083-0000779fd2ac.html#axzz2I0EiomcK (Accessed December 12, 2012).

The final answer to China's Arctic ambitions may come down to simple national pride. When other powerful countries have a stake in the Arctic, it would only be natural for China to seek equal status. Xu Yuanyuan, a Chinese professor at Qingdao University's School of Economics may have summed up China's true intentions:

> What, after all, is so alluring about the ice-in-the sky, snow-on-the-earth Arctic that it makes the three great and powerful countries the U.S., Canada, and Russia contentious to the point that they don't know what to do? After reading through many materials we have discovered [the reasons]: resources, sea routes, and strategic significance. These three resplendent jewels attract covetous stares from the three great and powerful countries.[13]

Just as China's ratification of UNCLOS could be a double-edge sword for them, so could the same situation apply for the U.S.'s status as non-party status to UNCLOS. China will give the U.S. little attention when being preached at to support international law.

Canada

Canada has always seen itself as a Northern nation, and the Arctic has been fundamental to that heritage. Like other Arctic nations, Canada clearly understands that changes in the region are creating both opportunities and challenges.

Canada has taken a keen interest in the reduction of summer ice. Understandably so, Canada's national interests differ from the U.S. The cornerstone of Canada's Arctic policy is laid out in its 2009 *Canada's Northern Strategy*. Listed in this document are Canada's four priorities of importance in the Arctic: exercising Arctic sovereignty, promoting social and economic development, protecting Canada's environmental

---

[13] David Curtis Wright, Canadian Defence and Foreign Affairs Institute., and Canadian Electronic Library (Firm), "*The Panda Bear Readies to Meet the Polar Bear China and Canada's Arctic Sovereignty Challenge*," Canadian Defence & Foreign Affairs Institute, p. 4

heritage, and improving and devolving Northern governance. Canada is looking to resolve international boundary disputes and gain recognition of its extended continental shelf. To this end, Canada has commissioned a major hydrographic exploration to map the ocean floor in support of a submission to the CLCS. Having ratified UNCLOS in 2003, Canada has ten years to submit such a claim and plans to do so in 2013. One of the key tenets of the strategy is decentralizing control from Ottawa and allowing for governments take a greater role in governance. To highlight the importance of including the indigenous populations in Arctic decision making, Prime Minister Stephen Harper announced that Inuit leader Leona Aglukkaq will serve as chair of the Arctic Council when Canada takes over in May 2013.[14] Canada asserted to build six to eight Arctic offshore vessels, a heavy conventional icebreaker for the Coast Guard, and develop an Arctic port in Nanisivik. To date, the progress for any of these initiatives has been slow or undetermined. Canada successfully completed an exercise called Operation NANOOK 2012 aimed at testing a Canadian whole of government approach to the defense of, and sovereignty over Canada's Arctic region.

Perhaps the most contentious sovereignty issue for Canada involves the United States. As the ice melts and transiting the Northwest Passage becomes more of a reality, Canada is embroiled with the United States on the legal classification of the waterway. Because it cuts through the archipelago of its many islands, Canada sees the Northwest Passage as passing through its internal waters. Thus, a transiting vessel would need Canadian permission and would be required to conform to all applicable Canadian laws. Canada could also levy large pilotage fees if it desired. The United States views the

---

[14] Prime Minister of Canada, "PM Appoints Canada's Chair of the Arctic Council," *Government of Canada*, http://www.pm.gc.ca/eng/media.asp?category=1&featureId=6&pageId=26&id=4973 (Accessed January 03, 2013).

Northwest Passage as an international strait, and that the right of transit passage should apply. In 1985, *USCGC POLAR SEA* sparked a political firestorm by transiting through the Northwest Passage without Canadian permission. Prime Minister Harper referred to the U.S. – Canadian disagreement of the Northwest Passes as a managed disagreement. The second dispute between Canada and the U.S. involves an EEZ boundary overlap between Alaska and the Yukon in the Beaufort Sea. In 2006 the discovery of an oil reserve with a possible 250 million barrels has drawn additional attention on the area.[15] Both U.S. and Canadian oil companies desire to begin exploratory drilling but cannot while an active disagreement on sovereignty exists.

The United States and Canada do share a common concern for Arctic security, a trend that has continued since the onset of the Cold War. In December of 2012, NORAD, USNORTHCOM, and Canadian Joint Operations Command (CJOC) signed the Framework for Arctic Cooperation that promotes "enhanced military cooperation in the preparation for and the conduct of defense, security, and safety operations in the Arctic."[16] This agreement is intended to be the first of several, and lays out plans for increased information sharing, domain awareness, and capability development.

Europe

Five of the eight Arctic nations are in Europe and all are ratified members of UNCLOS: Norway, Sweden, Iceland, Finland, and Denmark (by way of Greenland). Perhaps the two most important European Arctic nations are Denmark and Norway as

---

[15] Michael Byers, "Cold Peace: Arctic Cooperation and Canadian Foreign Policy," *International Journal* 65, no. 4 (Autumn 2010): 906.
[16] North American Aerospace Defense Command, United States Northern Command, and Canadian Joint Operations Command, *Framework for Arctic Cooperation* (Colorado Springs, CO: United States Northern Command, 2012), p. 2.

they are the only two states with Arctic coastlines (Iceland, Finland, and Sweden do not have coastlines north of the Arctic Circle).

Norway is a major exporter of oil and natural gas from sites in the North Sea. According to U.S. sources, Norway was the "second largest exporter of natural gas in the world after Russia, and the seventh largest exporter of oil," in 2011 and that accounted for nearly 50 percent of Norway's total GDP.[17] Western Europe is particularly dependent on Norway for much of its oil imports. With oil production already peaked in the North Sea, Norway began its first oil production in the Barents Sea in 2012 (north of the Arctic Circle). However, this move has drawn criticism from the European Union (E.U.) who has called for a moratorium on Arctic drilling, claiming that safety standards have not yet caught up with emerging technology. The energy sector is vital to Norway's economy and since they are not a member of the E.U., they are not bound by E.U. economic or governance policies. The U.S. is tied to Norway's success as U.S. companies invested over $29 billion in the oil and natural gas sector in 2011.[18]

Norway has made it clear that the E.U. has no jurisdiction over Norwegian affairs. But, this issue may heat up as Norway begins to extend its oil and natural production further into the Arctic. Bilateral talks are always an option, but since Norway and the E.U. are both party to UNCLOS, UNCLOS' standing dispute resolution mechanisms could be the path of choice for the two.

Greenland is an autonomous country within the Kingdom of Denmark. In 2009, the government of Denmark increased Greenland's right to self-governance through the *Greenland Self-Government Act of 2009*, including authority to decide how resources

---

[17] United States Energy Information Administration, "Norway," United States Department of Energy, http://www.eia.gov/countries/cab.cfm?fips=NO, (Accessed March 28, 2013).
[18] Ibid.

would be used (although Denmark still profits from Greenland's resources). The oil industry has never been a boon for Greenland, but the discovery of rare minerals on its north slope may be a valuable resource in the future.

Greenland has a long-standing dispute with Canada over Hans Island, a small uninhabited island that is in the center of the Nares Strait between Canada and Greenland. The island itself poses no real economic profit but could become politically important to both countries as this waterway begins a route into the Northwest Passage.

The real issue for Greenland will be its interpretation of its extended continental shelf (ECS) into the Arctic. In 2006, Denmark commissioned a major exploration to determine the ECS around Greenland's northern slope. Denmark is attempting to claim that Greenland's ECS extends all the way to the North Pole and encompasses much of the area on the Lomonosov Ridge that Russia has attempted to claim. If this were to become reality, Denmark could lay claim to the vast amount of natural resources that the region may hold. Denmark has not formally submitted a claim to the Commission on the Limits of the Continental Shelf (CLCS) but it is expected to by the end of 2014.

Greenland may become an important Arctic state for transportation and resources in the near future. As identified, Greenland has several high-profile disputes with other Arctic nations that need to be resolved. Denmark, a ratified member of UNCLOS, has identified the Arctic challenges it faces in its *Kingdom of Denmark Strategy for the Arctic 2011 – 2020*. Just like the language in *NSPD-66/HSPD-25*, Denmark reaffirmed its commitment to solving its disputes in a peaceful manner and identified UNCLOS as the primary framework for doing so.

CHAPTER 7: CONCLUSION AND RECOMMENDATIONS

The melting of the Arctic ice is no longer a matter of if; it is a matter of when. The United States, along with the international community, now faces important decisions that could shape the world's economic stability for the twenty-first century. The political environment of the Arctic is uncertain, but the Arctic Council and the United Nations offer a framework for peaceful international cooperation. Will it be a place of high tensions and military confrontation as countries race to claim the potential resource treasures that the Arctic holds? Will the Arctic serve as an example of respect for international law that works in the best interests of all countries? The United Nations Convention on the Law of the Sea treaty could be the cornerstone for creating those answers.

The United States has been described as a reluctant Arctic nation, with little interest in taking a leadership role in the region. By all accounts, that policy will not work for much longer. The 2010 *National Security Strategy* stated that the Arctic is key area to the advancement of U.S. interests. Future domestic and international security concerns will be met by building "new and deeper partnerships in every region, and strengthening international standards and institutions."[1] Becoming party to UNCLOS is clearly in line with, and supportive of, the national policy outlined in the *NSS*. As activity in the Arctic dramatically increases, the United States will be forced to take serious steps to include the Arctic as an integral element of two of its enduring national interests: homeland security and economic prosperity. To that end, the United States

---

[1] U.S. President, *National Security Strategy* (Washington, DC: Government Printing Office, 2010).

should become a fully ratified party to the United Nations Convention on the Law of the Sea, as one of the foundations of the nation's Arctic national strategy.

As its critics have noted, the treaty is not perfect, yet it is a necessary and worthwhile means to support national goals, while participating (and is some cases leading) in the international conversation on the Arctic. Because of this self-imposed exile from the international community, the U.S. may find itself on the outside looking in when other states are included in the creation of Arctic and international maritime policy. Those same states may have interests and agendas that are detrimental to U.S. interests. U.S. military, political, and business leaders are nearly unanimous in support for the immediate ratification of the treaty. They believe ratification of UNCLOS guarantees that the United States will continue to have a voice in the creation of international policies and in the protection of its national interests. The Arctic views of the United States are currently heard in forums such as the Arctic Council and the United Nations, but without the U.S. becoming a party to the convention, there is no guarantee that this will continue. This is particularly important for the Commission on the Limits on the Continental Shelf (CLCS) and the International Seabed Authority (ISA) as the United States is guaranteed permanent seats on both bodies with the ratification of UNCLOS. Currently, the U.S. has only observer status to the ISA and has no connection to the CLCS. Ratification of UNCLOS guarantees that the United States will continue to have a voice in the creation of international policies and in the protection of its national interests.

Codified freedom of navigation rights is undoubtedly in the best interest of the U.S. UNCLOS provides a basis for the military to operate in strategic international straits like the Strait of Malacca and Strait of Hormuz. In the near future, this freedom of navigation could become equally (perhaps even more) important in the Northwest or Northeast Passage in the Arctic.

If the U.S. becomes party to the Convention, it is guaranteed a permanent seat on the ISA. The U.S. is the only country in the world that can make such a boast. As the ISA requires consensus to act, with a permanent seat the U.S. would have the ability to veto any actions of the ISA that it deemed not in its national interest. A common fear about the ISA is that the U.S. would be required to pay royalties to an international body that would be redistributed to the world. While this is true to a certain extent, the amount is modest compared to the vast amount of money that U.S. companies could make in undersea mining and drilling enterprises. To take the ratification argument even further, U.S. energy companies were part of the U.S. negotiating team during the creation of the Implementation Agreement. The creation of the 1994 Implementation Agreement of Part XI was a political victory for the United States as the "international community went back to the drawing board and gutted the offending section of the treaty."[2] In essence, the United Nations re-wrote the treaty to incorporate and satisfy all the demands of the United States.

Perhaps the biggest incentive for the U.S. to become party to the convention is the potential for energy access and the effect this can have on the economy. The U.S. Geological Survey in 2008 claimed that there may be untold natural resources in the Arctic and that much of the usable oil could be in area that could be claimed by the U.S. With the benefit of a relatively shallow shelf of the Alaskan coast, the U.S. could have its

---

[2] Rob Huebert and University of Calgary, School of Public Policy., "*United States Arctic Policy the Reluctant Arctic Power,*" University of Calgary, School of Public Policy, p. 15.

sovereignty extended to almost 600 nautical miles offshore. The problem is that the United States cannot even make a continental shelf claim to the CLCS without being party to the convention. Critics will claim that the U.S. has rights to these waters on its continental shelf without the need for a treaty. The reality is that business will not invest the capital necessary to operate in the Arctic beyond the 200 nautical mile limits without the absolute certainty of law.

The Arctic is clearly changing. Whether or not the U.S. is prepared for those changes is less clear. Creating adequate policy and capabilities to operate in the Arctic will take time. Ratification of UNCLOS and utilization of that framework as a primary building block for Arctic strategy is vital to the U.S. The time for action is now because strategic considerations and national interests brought about by changes in the Arctic now require the United States to ratify the United Nations Convention on the Law of the Sea (UNCLOS).

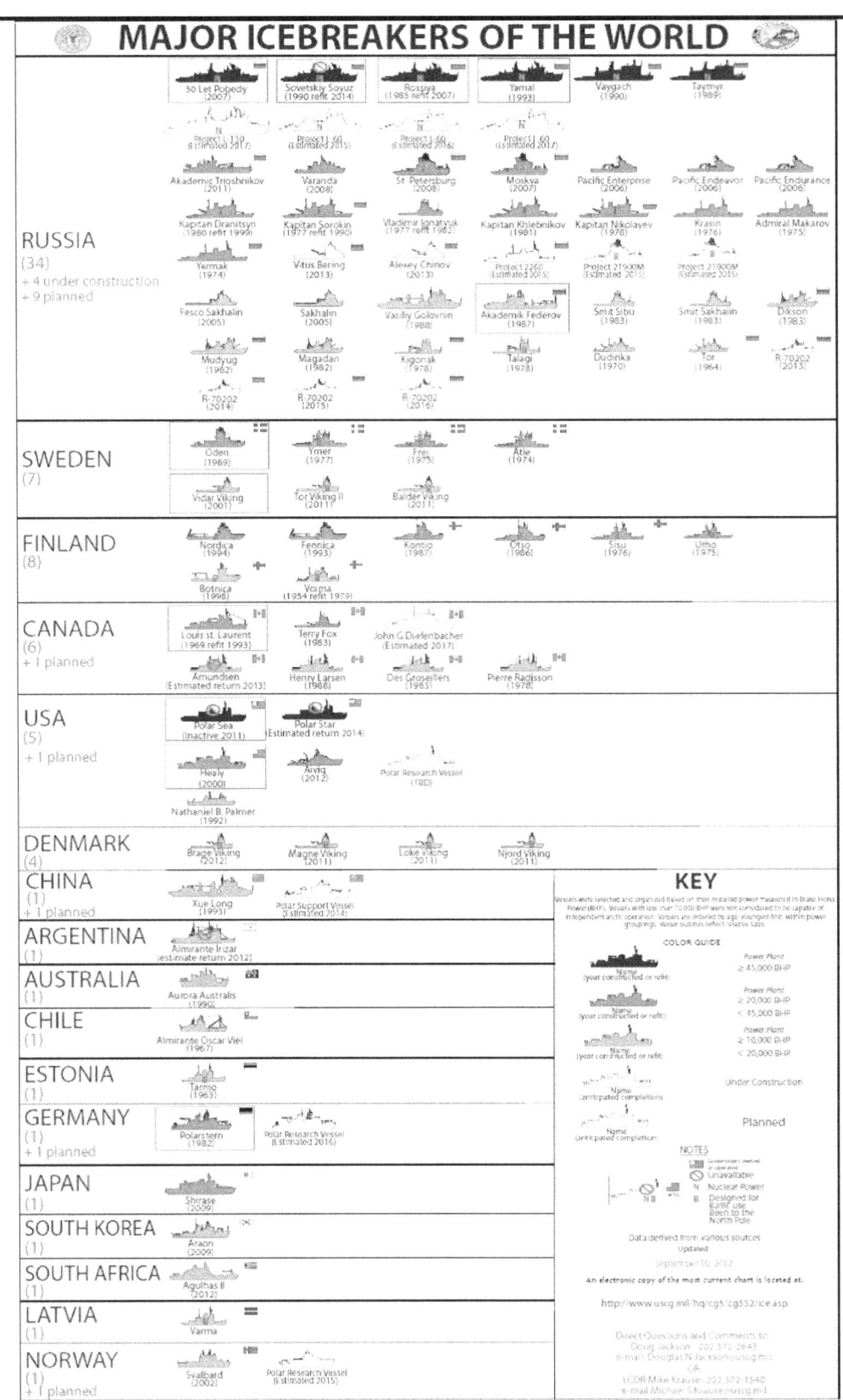

Developed and maintained by USCG Office of Waterways and Ocean Policy (CG-WWM)

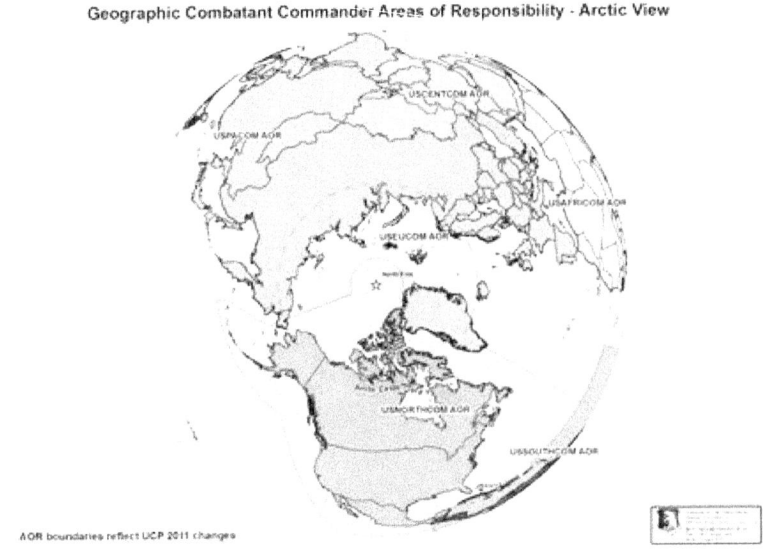

Source: Department of Defense Unified Command Plan 2011

BIBLIOGRAPHY

ABS Consulting. *United States Coast Guard High Latitude Regions Mission Analysis Capstone Summary*. Arlington, VA: ABS Consulting, 2010.

American Petroleum Institute, Chamber of Shipping of America, Financial Service Roundtable, International Association of Drilling Contractors, Marine Retailers Association of the Americas, National Association of Manufacturers National Marine Manufacturers Association, North American Submarine Cable Association, RARE, The Association for Rare Earth, TechAmerica Telecommunications Industry Association, U.S. Chamber of Commerce. Letter to Senators John Kerry and Richard Lugar supporting ratification of UNCLOS dated June 13, 2012, *ratifythetreatynow.org*. http://ratifythetreatynow.org/sites/default/files/pdf/Biz%20Support%20for%20Law%20of%20Sea%206-13-12.pdf. (Accessed January 21, 2013).

Arctic Portal. "Shipping." Nordurslodagattin, Akureyri, Iceland. http://www.arcticportal.org (Accessed September 26, 2012).

Bandow, Doug. "The Law of the Sea Treaty: Inconsistent with American Interests." *CATO Institute*. http://www.cato.org/publications/congressional-testimony/law-sea-treaty-inconsistent-american-interests (Accessed December 09, 2012).

Baumgartner, William D. "UNCLOS Needed for America's Security." *Texas Review of Law & Politics* 12, no. 2 (Spring 2008): 445-51,

Berkman, Paul Arthur and Royal United Services Institute for Defence and Security Studies. *Environmental Security in the Arctic Ocean: Promoting Co-Operation and Preventing Conflict*. Abingdon, UK: Published on behalf of The Royal United Services Institute for Defence and Security Studies by Routledge Journals, 2010.

Bert, Melissa, John Chaddic, and Brian D. Perry. "The Arctic in Transition-A Call to Action." *Journal of Maritime Law and Commerce* 40, no. 4 (Oct 2009): 481-509.

Borgerson, Scott G. "Arctic Meltdown: The Economic and Security Implications of Global Warming." *Foreign Affairs*, vol. 87 issue 2, (March/April 2008).

———. "The National Interests and the Law of the Sea", *Council on Foreign Relations: Council Special Report no. 46*, (May 2009).

British Broadcasting Company. "Russia Plants Flag under North Pole." *BBC News*, August 02, 2007. http://news.bbc.co.uk/2/hi/europe/6927395.stm. (Accessed January 12, 2013).

Bush, George H. W. "Letter to Congressional Leaders Transmitting a Report on Polar Icebreaker Requirements December 21, 1990." *American Reference Library - Primary Source Documents* (2001): p. 1, http://ezproxy6.ndu.edu/login?url=http://search.ebscohost.com/login.aspx?direct=true&db=mth&AN=32383016&site=ehost-live&scope=site. (Accessed January 19, 2013).

Byers, Michael. "*Who Owns the Arctic? Understanding Sovereignty Disputes in the North*." Vancouver: Douglas & McIntyre, 2009.

———. "Cold Peace: Arctic Cooperation and Canadian Foreign Policy." *International Journal* 65, no. 4 (autumn 2010): 899-912,

Canadian Press. "Summer Takes Unprecedented Toll on Arctic Ice." *Canada.com*, September 19, 2012. http://o.canada.com/2012/09/19/summer-takes-unprecedented-toll-on-arctic-ice-prompting-global-warming-fears/ (Accessed September 30, 2012).

Coffey, Luke. "Arctic Region: U.S. Policy on Arctic Security." *Heritage Foundation*. http://www.heritage.org/research/reports/2012/08/arctic-security-five-principles-that-should-guide-us-policy?rel=Arctic. (Accessed September 13, 2012).

Cohen, Richard E. and Peter Bell. "Congressional Insiders Poll: Do You Think it's been proven beyond a Reasonable Doubt that the Earth is warming because of Manmade Problems?" *National Journal*, (February 3, 2007). p. 6.

Cohen, Ariel. "From Russian Competition to Natural Resources Access: Recasting U.S. Arctic Policy", *Heritage Foundation*, 2010, http://www.heritage.org/research/reports/2010/06/from-russian-competition-to-natural-resources-access-recasting-us-arctic-policy. (Accessed January 10, 2013).

———. "Russia in the Arctic: Challenges to U.S. Energy and Geopolitics in the High North," in *Russia in the Arctic*, 1-42. Carlisle: Strategic Studies Institute, 2011.

Colman, Zack. "Republican Senator Says Sea Treaty might pass after Election." *The Hill*, August 17, 1012.

Congressional Research Service. *Coast Guard Polar Icebreaker Modernization Background, Issues, and Options for Congress*. Library of Congress. Washington, D.C.: Government Printing Office, 2008.

———. *Changes in the Arctic Background and Issues for Congress*. Library of Congress. Washington, D.C: Government Printing Office, 2011.

Conway, James T., Gary Roughead, and Thad W. Allen. "A Cooperative Strategy for 21st Century Seapower." *Naval War College Review* 61, no. 1 (Winter 2008): 6-19.

Demer, Lisa. "Shell Gambles Billions in Arctic Alaska Push." *Anchorage Daily News* December 04, 2011.

Department of Defense. United States Navy. *Navy Strategic Objectives for the Arctic*. Washington, DC: Department of Defense, May 21, 2010.

———. United States Navy. *Navy Arctic Roadmap*. Washington, DC: Department of Defense, November 10, 2009.

———. *Sustaining U.S. Global Leadership: Priorities for 21st Century Defense*. Washington, DC: Department of Defense. January 2012.

———. U.S. Navy. *Task Force Climate Change Charter*. Washington, DC: Department of Defense, October 30, 2009.

———. *Quadrennial Defense Review Report*. Washington, DC: Department of Defense. February 2010.

Departments of Defense and Homeland Security. U.S. Navy/U.S. Marine Corps, U.S. Coast Guard. *Naval Operations Concept*. Washington, DC: Department of Defense. 2010.

Department of Homeland Security. United States Coast Guard. "Coast Guard Completes Arctic Shield 2012." United States Coast Guard. http://www.uscgnews.com/go/doc/4007/1594651/Imagery-Available-Coast-Guard-completes-Arctic-Shield-2012 (Accessed November 11, 2012).

———. *United States Coast Guard 2012 Posture Statement*. Washington, DC: U.S. Coast Guard, February 2012. http://www.uscg.mil/history/allen/docs/CGFY2010PostureStatement.pdf

Dodds, Klaus. "Flag Planting and Finger Pointing: The Law of the Sea, the Arctic and the Political Geographies of the Outer Continental Shelf." *Political Geography* 29, no. 2 (2010).

Donohue, Thomas J., "Before the Committee on Foreign Relations of the United States Senate - Testimony of Thomas J. Donohue, President and Chief Executive Officer, U.S. Chamber of Commerce", June 28, 2012, http://www.foreign.senate.gov/imo/media/doc/Donohue%20Testimony.pdf, p. 2, (Accessed January 21, 2013).

Donohue, Thomas J. Letter to U.S. Senate expressing support for UNCLOS dated July 27, 2012, *ratifythetreatynow.org*, http://ratifythetreatynow.org/sites/default/files/pdf/Chamber%20of%20Commerce%20%2807-27-12%29.pdf, (Accessed January 21, 2013).

Emmerson, Charles. *The Future History of the Arctic*. New York: Public Affairs, 2010.

Gaffney Jr., Frank J. "U.N.'s Larger Role in UNCLOS is Bad for American Interests." *Texas Review of Law & Politics 12*, no. 2 (2008), p. 469-476.

Garrett, Jeffrey M. "Enduring Arctic Reluctance." *sldinfo.com*. July 07, 2011 http://www.sldinfo.com/ending-reluctance/ (Accessed November 12, 2012).

Gautier, Catherine. Oil, Water and Climate: An Introduction. New York, NY: Cambridge University Press, 2008.

Geology.com. "Oil and Natural Gas Resources of the Arctic." http://geology.com/articles/arctic-oil-and-gas/ (Accessed January 02, 2013).

Grant, Shelagh D. *Polar Imperative: A History of Arctic Sovereignty in North America*. Vancouver; Berkeley [Calif.]; [Berkeley, Calif.]: Douglas & McIntyre; Distributed in the U.S. by Publishers Group West, 2010.

———. *Sovereignty or Security? : Government Policy in the Canadian North*, 1936-1950. Vancouver: University of British Columbia Press, 1988.

Greenert, Jonathan. "UNCLOS and U.S. Freedom of Navigation." *Hampton Roads International Security Quarterly*. (Jul 1, 2012): p. 37.

Groves, Steven. "Costs of UNCLOS Accession." *Hampton Roads International Security Quarterly*. (Jul 1, 2012): p. 63.

Howard, Roger. *The Arctic Gold Rush: The New Race for Tomorrow's Natural Resources*. London; New York: Continuum, 2009.

Huebert, Rob and University of Calgary. School of Public Policy. "United States Arctic Policy the Reluctant Arctic Power." University of Calgary, School of Public Policy.

Holmes, Kim R. "What's the Big Idea: U.N. Sea Treaty Still a Bad Deal for U.S." *The Washington Times*, July 14, 2011. p. 4. http://www.washingtontimes.com/news/2011/jul/13/holmes-un-sea-treaty-still-a-bad-deal-for-us/ (Accessed January 12, 2013).

Ilulissat Conference. The Ilulissat Declaration. Ilulissat, Greenland: 2008.

Jacoby, Charles H. "UNCLOS Vital to Arctic Cooperative Security." *Hampton Roads International Security Quarterly* (Jul 1, 2012): 45.

Jakobson, Linda and Stockholm International Peace Research Institute. "China Prepares for an Ice-Free Arctic." *Stockholm International Peace Research Institute (SIPRI)*. No. 2010/2. March 2010.

Johansen, Bruce E. *Global Warming 101*. Westport, CT: Greenwood Press, 2008.

Karl, Thomas and Kevin Trenberth. "Modern Global Climate Change." *Science* Vol 202, no. 56512003. p. 1719.

Kissinger, Henry. "National Security Decision Memorandum 144." *Federation of American Scientists*. http://www.fas.org/irp/offdocs/nsdm-nixon/nsdm-144.pdf (Accessed October 4, 2012).

Kissinger, Henry, and George Shultz, James Baker III, Colin Powell, and Condoleezza Rice, "Time to Join the Law of the Sea Treaty," *online.wsj.com*, May 20, 2012. http://online.wsj.com/article/SB10001424052702303674004577434770851478912.html (Accessed December 12, 2012).

Kramer, Andrew E. "Warming Revives Dream of Sea Route in Russian Arctic." *New York Times*, October 17, 2011. http://www.nytimes.com/2011/10/18/business/global/warming-revives-old-dream-of-sea-route-in-russian-arctic.html?pagewanted=all&_r=0 (Accessed December 02, 2012).

Lugar, Richard. "Failure to Ratify UNCLOS A Position of Self-Imposed Weakness." *Hampton Roads International Security Quarterly* (Jul 1, 2012): 13.

Matthew, Richard A. "Is Climate Change a National Security Issue?" *Issues in Science and Technology* 27, no. 3 (Spring 2011, 2011): 49-60.

Moore, John N. "UNCLOS Key to Increasing Navigational Freedom." *Texas Review of Law & Politics* 12, no. 2 (Spring 2008, 2008): 459-67.

Moskvitch, Katia. "Russia to Build Biggest Nuclear-Powered Icebreaker." *British Broadcasting Company*. http://www.bbc.com/news/technology-19576266 (Accessed September 17, 2012).

National Aeronautical and Space Administration. "Arctic Sea Ice Hits Smallest Extents in Satellite Era." *NASA*. http://www.nasa.gov/topics/earth/features/2012-seaicemin.html (Accessed January 13, 2013).

National Oceans and Atmospheric Administration. "Climate Watch Magazine: Hottest Month Ever Recorded." *NOAA*. http://www.climatewatch.noaa.gov/image/2012/july-2012-hottest-month-on-record. (Accessed September 17, 2012).

Negroponte, John. "Joining UNCLOS Now: Nine Reasons why it is Imperative for the United States." *Hampton Roads International Security Quarterly* (Jul 1, 2012): 59.

Nordquist, Myron H. *Legal and Scientific Aspects of Continental Shelf Limits*. Reykjavik, Iceland ed. Leiden; Boston: Martinus Nijhoff, 2004.

Nordquist, Myron H. et al. *Oceans Policy: New Institutions, Challenges, and Opportunities*. Vol. 22nd annual seminar. The Hauge, Netherlands: Kluwer Law International, 1999.

Nordquist, Myron H., John Norton Moore, and University of Virginia. Center for Oceans Law and Policy. *Current Maritime Issues and the International Maritime Organization*. The Hague; Boston: Martinus Nijhoff Publishers, 1999.

Nordquist, Myron H. and University of Virginia. Center for Oceans Law and Policy. Conference. *The Law of the Sea Convention: US Accession and Globalization*. The Hague; Boston: Martinus Nijhoff Publishers, 2012.

Nordquist, Myron H., Moore, John N., Skaridov, Alexander S. and University of Virginia. *International Energy Policy, the Arctic, and the Law of the Sea*. Boston: Martinus Nijhoff Publishers, 2005.

North American Aerospace Defense Command and United States Northern Command. Arctic Collaborative Workshop: Arctic Oil Spill & Mass Rescue Operation - Tabletop Exercise, After Action Report. Colorado Springs, CO: United States Northern Command, 2012.

North American Aerospace Defense Command, United States Northern Command, and Canadian Joint Operations Command. Framework for Arctic Cooperation. Colorado Springs, CO: United States Northern Command, 2012.

Panetta, Leon. "UNCLOS Accession would Strengthen US Global Position." *Hampton Roads International Security Quarterly* (Jul 1, 2012): 23.

Papp, Robert J. "Enhancing Coast Guard Operations through UNCLOS Accession." *Hampton Roads International Security Quarterly* (Jul 1, 2012): 47.

Payne, Amy. "Republican Opposition Downs U.N. Disability Treaty." *USA Today*. http://www.usatoday.com/story/news/politics/2012/12/04/disability-united-nations-treaty-senate-dole/1745679/ (Accessed February 02, 2013).

Prime Minister of Canada. "PM Appoints Canada's Chair of the Arctic Council." *Government of Canada*. http://www.pm.gc.ca/eng/media.asp?category=1&featureId=6&pageId=26&id=4973 (Accessed January 03, 2013).

Revkin, Andrew C. "Experts Urge U.S. to Increase Icebreaker Fleet in Arctic Waters." *New York Times,* August 17, 2008: http://www.nytimes.com/2008/08/17/world/europe/17arctic.html?_r=0 (Accessed November 16, 2012).

Roi, Michael L. "Russia: The Greatest Arctic Power?" *Journal of Slavic Military Studies* 23, no. 4 (Oct, 2010): 551-73.

Rosett, Claudia. "Magic with U.S. Money for the United Nations." *Forbes Inc.* http://www.forbes.com/sites/claudiarosett/2011/04/08/magic-with-u-s-money-for-the-united-nations/ (Accessed December 10, 2012).

Ruel, Geneviève King. "The (Arctic) show must go on." *International Journal* 66, no. 4 (autumn 2011, 2011): 825-33,

Sacco, Michael. Letter to Secretary of State Hillary Clinton dated September 21, 2011, *ratifythetreatynow.org*, http://ratifythetreatynow.org/sites/default/files/pdf/Letter-Sacco-September-2011.pdf, (Accessed January 21, 2013).

Schiller, Bill and Toronto Star. "China Warming Up to be an Arctic Player." *The Toronto Star*, March 1, 2010. (Accessed February 02, 2013).

Schofield, Clive and Tavis Potts. "Across the Top of the World? Emerging Arctic Navigational Opportunities and Arctic Governance." *Carbon & Climate Law Review* 3, no. 4 (12/30, 2009): 472-82.

Schultz, George P., "A Changed World", Lecture – Library of Congress, Washington, DC, March 22, 2004.

Senate Armed Services Committee. "Jeanne J. Kirkpatrick testimony before the Senate Armed Services Committee April 8, 2004". http://www.armed-services.senate.gov/statemnt/2004/April/Kirkpatrick.pdf. (Accessed January 19, 2013).

Senate Foreign Relations Committee. "Accession to the 1982 Law of the Sea Convention and Ratification of the 1994 Agreement Amending Part XI of the Law of the Sea Convention," http://www.foreign.senate.gov/imo/media/doc/REVISED_Secretary_Clinton_Testimony.pdf. (Accessed March 10, 2013).

Snow, Nick. "BSEE Approves Shell's Chukchi Sea Oil Spill Response Plan." *Oil & Gas Journal*, Feb 27, 2012. p. 14.

Spears, Joseph. China and the Arctic: The Awakening Snow Dragon. ed. Vol. 9. Washington, DC: *The Jamestown Foundation*, 2009. http://www.jamestown.org/programs/chinabrief/single/?tx_ttnews%5Btt_news%5D=34725&cHash=9638471049 (Accessed January 14, 2013).

Spring, Baker. "All Conservatives should Oppose UNCLOS." *Texas Review of Law & Politics* 12, no. 2 (Spring 2008): 453-7.

Stringer, David. "UK Lawmakers Seek Moratorium on Arctic Drilling." *Seattle Times*, September 19, 2012. (Accessed October 04, 2012).

Tan, Wei-en and Yu-tai Tsai. "After the Ice Melts: Conflict Resolution and the International Scramble for Natural Resources in the Arctic Circle." *Journal of Politics and Law* 3, no. 1 (Mar 2010): 91-99.

The United States Global Change Research Program. Committee on Environment and Natural Resources. http://downloads.globalchange.gov/usimpacts/pdfs/southeast.pdf (Accessed September 07, 2012).

Tillerson, Rex W. Law of the Sea Treaty. Letter to Senators John Kerry and Richard Lugar stating Exxon's support for ratification of the Law of the Sea Treaty 2012. http://ratifythetreatynow.org/sites/default/files/pdf/ExxonMobil%20%2806-08-12%29.PDF (Accessed January 21, 2013).

Timmons, Jay. "UNCLOS Critical for US Manufacturing Competitiveness." *Hampton Roads International Security Quarterly* (Jul 1, 2012): p. 103.

Titley, David W. and Courtney C. Saint John. "Arctic Security Considerations and the U.S. Navy's Arctic Roadmap." *Arctic Security in an Age of Climate Change Arctic Security in an Age of Climate Change.* New York: Cambridge University Press, (2011): 267-80.

Traner, Helena. "Resolving Arctic Sovereignty from a Scandinavian Perspective." *Case Western Reserve Journal of International Law* 44, no. 1/2 (2011): 497-525.

Truver, Scott C. "UNCLOS Mythbusters." United States Naval Institute. Proceedings vol. 133, no. 7 (Jul 2007): 52-53.

United Nations. "Submissions to the Commission on the Limits of the Continental Shelf". *www.un.org*, http://www.un.org/Depts/los/clcs_new/commission_submissions.htm. (Accessed October 19, 2012).

United Nations. "United Nations Convention on the Law of the Sea." *www.un.org*. http://www.un.org/Depts/los/convention_agreements/texts/unclos/closindx.htm. (Accessed August 26, 2012).

United Nations Division for Ocean Affairs and the Law of the Sea. "*Commission on the Limits of the Continental Shelf.*" United Nations. http://www.un.org/Depts/los/clcs_new/clcs_home.htm (Accessed October 19, 2012).

United Nations Environment Programme (UNEP). "*Background to UNCLOS.*" UNEP/GRID-Arendal. http://continentalshelf.org/about/1143.aspx (Accessed November 28, 2012).

United States Arctic Research Commission. "*Arctic Marine Shipping Assessment: Current Marine use and the AMSA Database.*" United States Arctic Research Commission. http://www.arctic.gov/publications/AMSA/current_marine_use.pdf (Accessed September 10, 2012).

United States Congress. United States Code Title 14 - Coast Guard. Vol. chapter 393, 1, 63 statute 495. Washington, DC: August 4, 1949.

———. Arctic Research and Policy Act of 1984 (Amended 1990). Vol. Public Law 98-373; Public Law 101-609. Washington, DC: 1984; 1990.

United States Energy Information Administration. "Russia." United States Department of Energy. http://www.eia.gov/countries/cab.cfm?fips=RS (Accessed December 22, 2012).

———. "Natural Gas Demand at Power Plants was high in summer 2012." United States Department of Energy. http://www.eia.gov/todayinenergy/detail.cfm?id=7870 (Accessed January 31, 2013).

———. "Norway." United States Department of Energy. http://www.eia.gov/countries/cab.cfm?fips=NO. (Accessed March 28, 2013).

———. "China." United States Department of Energy. http://www.eia.gov/countries/country-data.cfm?fips=CH&trk=c. (Accessed January 09, 2012).

———. "Arctic Oil and Natural Gas Resources." United States Department of Energy. http://www.eia.gov/oiaf/analysispaper/arctic/index.html#adcr (Accessed September 20, 2012).

United States Department of the Interior. U.S. Geological Survey. *Circum-Arctic Resource Appraisal: Estimates of Undiscovered Oil and Gas North of the Arctic Circle.* Open-file report, U.S. Geological Survey. Washington DC, 2008. http://pubs.usgs.gov/fs/2008/3049/fs2008-3049.pdf (Accessed September 10, 2012).

United States Northern Command. "Joint Task Force Alaska." United States Northern Command. http://www.northcom.mil/About/index.html#JTFAK (Accessed December 04, 2012).

United States. Congress. Senate. Committee on Appropriations. Subcommittee on the Department of Homeland Security. "Strategic Importance of the Arctic in U.S. Policy Hearing before a Subcommittee of the Committee on Appropriations, United States Senate, One Hundred Eleventh Congress, Special Hearing, August 20, 2009, Anchorage, AK." U.S. Government Printing Office.

United States. Congress. Senate. Committee on Commerce, Science, and Transportation. Subcommittee on Oceans, Atmosphere, Fisheries, and Coast Guard. "Defending U.S. Economic Interests in the Changing Arctic is there a Strategy? : Hearing before the Subcommittee on Oceans, Atmosphere, Fisheries, and Coast Guard of the Committee on Commerce, Science, and Transportation, United States Senate, One Hundred Twelfth Congress, First Session, July 27, 2011." U.S. Government Printing Office.

U.S., Canada, Denmark, Norway, Russia. *Ilulissat Declaration of 1997*. Ilulissat, Greenland: May 28, 2008.

U.S. President, *National Security Strategy*. Washington, DC: Government Printing Office, May 2010.

U.S. President. Proclamation. "National Security Presidential Directive 66/Homeland Security Presidential Directive 25: Arctic Region Policy." January 9, 2009.

———. Proclamation. "President Ronald Reagan's Statement on United States Participation in the Third United Nations Conference on the Law of the Sea." (January 29, 1982).

———. Proclamation. "President's Statement on Advancing U.S. Interests in the World's Oceans." (May 15, 2007). http://search.proquest.com.ezproxy6.ndu.edu/docview/190573909?accountid=12686 (Accessed December 16, 2012).

———. Proclamation. "Proclamation 2667: Policy of the United States with Respect to the Natural Resources of the Subsoil and Sea Bed of the Continental Shelf ." Federal Register 10, no. 12305 (September 28, 1945).

———. Proclamation. "Proclamation 2668: Policy of the United States with Respect to Coastal Fisheries in Certain Areas of the High Seas," Federal Register 10, no. 12304 (September 28, 1945).

———. Proclamation. "Statement on United States Oceans Policy." (March 10, 1983).

Voronov, Konstantin. "The Arctic Horizons of Russia's Strategy." *Russian Politics & Law* 50, no. 2 (Mar, 2012): 55-77.

Wade, Robert. "A Warmer Arctic Ocean Needs Shipping Rules." *Financial Times*. http://www.ft.com/intl/cms/s/0/1c415b68-c374-11dc-b083-0000779fd2ac.html#axzz2I0EiomcK (Accessed December 12, 2012).

Weigie, Gao. "Development Strategy of Chinese Shipping Company under the Multilateral Framework of WTO." *cosco.com*. http://www.cosco.com/en/pic/forum/654923323232.pdf (Accessed February 02, 2013).

Wong, Queenie. "New Study Estimates Vast Supplies of Arctic Oil, Gas." McClatchy Newspapers, July 24, 2008, 2008., http://www.mcclatchydc.com/2008/07/23/v-print/45349/new-study-estimates-vast-supplies.html (Accessed September 12, 2012).

Wright, David C., and Naval War College (U.S.). China Maritime Studies Institute. *The Dragon Eyes the Top of the World Arctic Policy Debate and Discussion in China*. Newport, RI: U.S. Navy War College, August 2011.

Wright, David C., and Canadian Defence and Foreign Affairs Institute. *The Panda Bear Readies to Meet the Polar Bear China and Canada's Arctic Sovereignty Challenge*. Calgary: Canadian Defence & Foreign Affairs Institute, 2011.

Zellen, Barry Scott. *Arctic Doom, Arctic Boom: The Geopolitics of Climate Change in the Arctic*. Santa Barbara, Calif.: Praeger, 2009.

Zysk, Katarzyna. "Russia's Arctic Strategy." *Joint Force Quarterly*, no. 57 (2010): 103-110.

VITA

A native of the Outer Banks of North Carolina, LCDR Gray is a 1995 graduate of the University of North Carolina at Wilmington and a 2005 graduate of the National Graduate School. Completing Coast Guard basic training in 1995, LCDR Gray served onboard USCGC DALLAS (WHEC 716) as a Seaman and USCGC DURABLE (WHEC 628) as a Quartermaster Second Class. Earning a commission from Coast Guard Officer Candidate School in Yorktown, VA in 1998, LCDR Gray served onboard several operational cutters: as Operations Officer onboard USCGC HORNBEAM (WLB 394), as First Lieutenant onboard USCGC DILIGENCE (WMEC 616), as Combat Information Center Officer onboard USS HAYLER (DD 997), as Executive Officer onboard USCGC SEQUOIA (WLB 215), and as Commanding Officer onboard USCGC MAPLE (WLB 207). LCDR Gray's shore tours include the Office of Defense Operations at Coast Guard Headquarters and Coast Guard Liaison Officer at U.S. Third Fleet. Upon graduation from the Joint Forces Staff College, LCDR Gray will be assigned to Coast Guard Atlantic Area as a Joint Operational Planner.

www.ingramcontent.com/pod-product-compliance
Lightning Source LLC
Chambersburg PA
CBHW081142170526
45165CB00008B/2770